P9-BIM-998

Range Science

NATURAL WORLD INFORMATION GUIDE SERIES

Series Editor: Russell Shank, University Librarian, Library Administrative Office, University of California, Los Angeles

Also in this series:

ENERGY STATISTICS—*Edited by Sarojini Balachandran*

The above series is part of the

GALE INFORMATION GUIDE LIBRARY

The Library consists of a number of separate series of guides covering major areas in the social sciences, humanities, and current affairs.

General Editor: Paul Wasserman, Professor and former Dean, School of Library and Information Services, University of Maryland

Managing Editor: Denise Allard Adzigian, Gale Research Company

Range Science

A GUIDE TO INFORMATION SOURCES

Volume 2 in the Natural World Information Guide Series

John F. Vallentine

Professor of Range Science
Brigham Young University
Provo, Utah

Phillip L. Sims

Research Leader, USDA
U.S. Southern Great Plains Field Station
Woodward, Oklahoma

Gale Research Company
Book Tower, Detroit, Michigan 48226

Library of Congress Cataloging in Publication Data

Vallentine, John F
Range science.

(Natural world information guide series ; v. 2)
(Gale information guide library)
Includes indexes.
1. Range management—Bibliography. 2. Rangelands—Bibliography. 3. Range management—Information services. 4. Rangelands—Information services. I. Sims, Phillip L., joint author. II. Title. III. Series.
Z5074.R27V33 [SF85] 016.33374 80-14361
ISBN 0-8103-1420-7

VITA

Dr. Vallentine, a native of Kansas, has spent more than twenty years in range science as teacher, educator, extension specialist, scientist, and author. He has authored and coauthored numerous publications, including the textbook, RANGE DEVELOPMENT AND IMPROVEMENTS, and the compilation, U.S.-CANADIAN RANGE MOVEMENT, 1935-1977: A SELECTED BIBLIOGRAPHY. He was formerly extension range specialist at Utah State University and subsequently held an extension-research position at the University of Nebraska North Platte Station. He is a graduate of Kansas State University (B.S.), Utah State University (M.S.), and Texas A&M University (Ph.D.).

Dr. Sims's accomplishments include authorship or coauthorship of numerous publications in range science and leadership in the thirty-year computer compilation of bibliographic citations in the Journal of Range Management. Before his present USDA research appointment in Oklahoma, the state of which he is a native, he was on the Colorado State University research and teaching faculty for more than ten years. Dr. Sims is a graduate of Oklahoma State University (B.S. and M.S.) and Utah State University (Ph.D.).

CONTENTS

Contents

PREFACE

Range science has existed for many years with compilations of information dating back a half century. During the more than thirty years of existence of the Society for Range Management, volumes of literature have been published regarding basic concepts, descriptions, developments, and management principles and practices in range science. With this literature explosion, one's imagination is flooded in an attempt to stay current, and only with computerized retrieval systems would it be possible to cope with the proliferation of information. Because of the diversity of sources and their locations, instructions are generally lacking for efficient use of the vast amount of published information in range science.

This guide provides a valuable key for today's range science information seeker --whether student, scientist, educator, manager, or administrator. This comprehensive guide, the first of its kind, pulls together primary sources of information in range science and identifies organizations and agencies from which assistance may be obtained, including searching procedures. The format presents range science--its status and development, literature searching, sources of information, selected literature, and a subject index. The simplicity in subject and topic organization makes it readily usable.

Henry A. Pearson
Chief Range Scientist
USDA, Forest Service
Southern Forest Expt. Sta.
Pineville, Louisiana 71360

ACKNOWLEDGMENTS

The authors wish to express their appreciation to D. Lynn Drawe, Richard D. Jensen (BYU Life Sciences Librarian), Henry A. Pearson, Rex D. Pieper, Russell Shank (UCLA Librarian), and David A. Smith for reviewing parts or all of the manuscript.

INTRODUCTION

It has been readily apparent that a guide to range science information has long been needed by library users and librarians alike. Equally apparent was that range science informational needs are not limited to library patrons. This encouraged the scope of this guide to be expanded to meet the needs of all people engaged in providing, preparing, and retrieving as well as using range science information. This guide is dedicated to the more effective use of available range science information sources.

The need for range science information by students, teachers, and research scientists is widely recognized; but the need for this information by livestock producers, big game managers, natural resource managers, range consultants, and agribusiness personnel is just as strong and continuing. It is also of use to conservationists generally, legislators, and land use planners. In fact, it is concluded that nobody could be excluded from eventually needing range science facts and information.

The functional needs for range science information are many and include the following:

1. Making administrative and managerial decisions
2. Preparing for research
3. Meeting preliminary requirements toward theses and dissertations
4. Preparing classroom and extension materials
5. Writing review articles, reports, term papers, books, and manuals
6. Developing environmental impact statements
7. Preparing range and ranch plans or consultant reports
8. Fulfilling personal interests in becoming informed
9. Meeting legislative and general public affairs needs

In addition to use as a range science reference manual, this guide has also been prepared for use in a graduate or undergraduate literature search seminar. An appendix section has been devoted to an outline and suggestions for adapting this guide to use as a syllabus for such a seminar.

Chapter 1

RANGE SCIENCE—ITS STATUS AND DEVELOPMENT

DESCRIPTION OF RANGE

RANGE: uncultivated grasslands, shrublands, or forested lands with an herbaceous and/or shrubby understory, particularly those producing forage for grazing or browsing by domestic and wild animals; includes lands with native vegetation cover and also lands naturally or artificially revegetated with native or adapted, introduced forage plant species not requiring periodic reestablishment and subsequently managed like native range.

Although the concept of range developed in western North America, it has now been applied to lands throughout the world. The term formerly connoted broad, open, unfenced, undeveloped lands supporting only native plants and receiving minimal management. However, this concept of range is no longer accurate or useful because of the intensification of management and the treatment and development practices now being widely applied. The land-based concept of range has evolved to a considerable degree to reflect the range ecosystem, an interacting system composed of both natural and introduced biotic and abiotic components including not only the vegetation and soil but also the associated atmosphere, water, and animals.

Rangelands are highly diversified and include natural grasslands, savannas, shrublands, most deserts, tundra, alpine communities, coastal marshes, and wet meadows.[1] Rangelands commonly have physical limitations that make them unsuited to cultivation; these factors may include rocky, shallow, sandy, or saline soils, low and erratic precipitation, steep or rough topography, poor drainage, or extreme temperatures. However, some arable lands may be left as rangeland by choice. Also, range may be derived by revegetating unproductive or eroded croplands and temporarily by logging and burning dense forests. Some rangelands may best be classified concurrently as cropland and forest land also, and shifts are continually being made between these land categories. Pasture, defined as lands producing plant materials harvested directly by grazing animals, includes range or rangelands but also cultivated perennial, temporary, and crop aftermath pasture.[2]

The use or uses made of range vary from area to area and over time. Although the production of forage for direct consumption by grazing or browsing animals is the single, most nearly universal characteristic of range, some ranges may remain ungrazed by livestock because of physical limitations, economics, or higher priorities being given to restrictive usage such as watershed or recreation. Most ranges, however, both privately and publicly owned, provide a combination of multiple uses and yield a variety of products.

Range in the United States exclusive of Hawaii and Alaska was inventoried in 1970 under the Forest-Range Environmental Study (FRES).[3] As defined in the study, and referred to therein as "forest-range," range included native and natural grasslands and shrublands as well as forested lands that do or can be made to produce vegetation grazeable by livestock. The study revealed that 63 percent of the United States or 1,201.6 million acres was range. Range comprised 83 percent of the West, 53 percent of the Plains, 41 percent of the Northeast, and 61 percent of the Southeast. By ownership, 31 percent of the U.S. range was federal and 69 percent nonfederal.

The FRES report further revealed that 43.5 percent of the United States was range actually grazed by livestock. The portion of the rangelands which was grazed by livestock in 1970 varied from a high of 86 percent in the Plains and 78 percent in the West to a low of 34 percent in the Northeast and 46 percent in the Southeast. While the estimated 213 million animal unit months of grazing realized by domestic livestock from range in 1970 was spread nationwide, 75 percent was furnished by the Plains and West. In addition range provided habitat and grazing capacity for more than 5 million big-game animals and countless numbers of other wildlife species.

The FRES study concluded that range livestock grazing annually provided animal output value of $1.7 billion (1970 basis) and direct employment of 196 million man-hours of work. Assuming that production agriculture generates total agribusiness employment at a 1:9 ratio and that 2,112 man-hours is equivalent to one man year, range livestock grazing can be credited with providing full-time employment for 7.2 million people. Range livestock grazing, of course, is only one of several economic products from rangelands.

Livestock grazing capacity requirements from range were predicted by FRES to be 320 million animal unit months by the year 2000. However, it was estimated that range in the United States had the potential of producing 566 million animal unit months by the year 2000 under intensive range and livestock management while still maintaining the environment and providing for multiple use.

RANGE SCIENCE AS A DISCIPLINE

RANGE SCIENCE: the organized body of knowledge upon which is based the practice of range management, defined as planning and directing the use of rangelands to obtain optimum, sustained returns based on the objectives of land ownership and society needs and desires generally.

Range science is a distinct discipline dealing with the protection, development, and utilization of range and associated resources. Since range science has a central objective of producing forage for consumption by domestic and wild herbivores, it is basically involved in the manipulation of the range plant-range soil-range animal complex of range ecosystems. However, grazeable forage as a principal range resource is associated with other range resources, these available from range in varying amounts and combinations over time and from one location to another. The latter include both tangible products and intangible values such as wildlife, water, landscape and natural beauty, recreational opportunities, minerals, wood products, germ plasm for domestication and breeding, and areas for the study of the ecology of natural ecosystems.[4]

Range science is a relatively new discipline and has evolved as a synthesis of other disciplines with which it still maintains common interests. Based on grazing or herbivory being its essential feature, range science has been described as a combination of biological, physical, and social science.[5] As a plant (biological) science, it is concerned with plant production and the response of plants to cropping (grazing) and is thus allied with crop science, plant ecology, and plant physiology. As an animal (biological) science, it deals with animal grazing capacities and responses of animals harvesting the range forage; it thereby relates to animal production (animal science), wildlife biology, and zoology.

Since climatic, topographic, edaphic, and hydrologic factors determine the productivity and kind and degree of use that can be made of range, range science as a physical science relates to meterology, climatology, hydrology, geology, and soil science. As a social science it is concerned with natural resource economics, ranch economics, sociology, and land use planning. Because of integrated uses of range resources, it has common interest with forest management, outdoor recreation, and watershed management.

The following subject guide shows the component parts of range science in outline form.[6]

SUBJECT GUIDE TO RANGE SCIENCE

1. Range plants and forages
   - A. Plant systematics
   - B. Forage value (including poisonous plants)
   - C. Plant genetics and breeding
   - D. Plant pathology and insects
   - E. Plant physiology

2. Range ecology
   - A. Climatic factors
   - B. Biotic factors
   - C. Edapic factors
   - D. Pyric factors
   - E. Autecology
   - F. Synecology

3. Range resources
   A. Range inventory
   B. Natural resources planning
   C. Public land administration

4. Grazing management
   A. Stocking rates
   B. Grazing seasons and systems
   C. Grazing habits and distribution

5. Range and ranch development
   A. Plant control
   B. Range revegetation
   C. Special treatments
   D. Fertilization
   E. Livestock handling facilities

6. Range livestock
   A. General management
   B. Nutrition and feeding
   C. Breeding and reproduction
   D. Veterinary medicine

7. Range wildlife
   A. Big game
   B. Upland game
   C. Other game
   D. Nongame

8. Range watersheds
   A. Water resources
   B. Watershed influences
   C. Hydrology and water yield
   D. Watershed restoration

9. Ranch economics
   A. Ranch planning and organization
   B. Business management
   C. Capital and credit
   D. Ranch values and appraisal
   E. Livestock marketing

10. Range education and research
   A. Range science philosophy
   B. Professional training
   C. Adult and youth education
   D. Research

DEVELOPMENT OF RANGE SCIENCE [7]

Although the pastoral art of grazing management had its origin in prehistory, the development of range management as a science did not begin until late in the nineteenth century. The term range has been used to mean natural pasture as early as 1846 in referring to "a common range for cattle" in Oregon,[8] and the term can probably be documented back some years before this. Range management as a profession, however, was first evidenced in the 1890s when the Division of Agrostology, later part of the Bureau of Plant Industry, began formulating and reporting the principles of range science in west Texas and in the Pacific northwest.

Range research prior to 1905 was mostly observational and exploratory. By 1910 eight state experiment stations and the U.S. Forest Service were experimentally investigating range problems. Although some formal grazing experiments had begun by 1910, the development of range research has approached the needed comprehensive program only since about 1935.

Range Management on the National Forests, a bulletin by Jardine and Anderson[9] published by USDA in 1919, might be considered the first successful attempt to compile in organized form the principles of range science. These principles were put in more comprehensive form in 1923 and 1928 by A.W. Samson,[10] who is sometimes referred to as the "father of range science." However, the most widely known textbook on range science and the one probably having the greatest impact on the formalization of range science was RANGE MANAGEMENT by Stoddart and Smith (1943).[11]

The first formal range management course was apparently offered in 1914 at Utah State College, Logan, but related courses were also being taught by 1920 at Montana State University, Missoula; Colorado A&M; University of Idaho; Washington State College; University of California; and the University of Minnesota. The first range management curriculum was begun at Montana State University in 1916, being followed the next year at the University of Idaho. The first doctoral degrees in range science were awarded by Texas A&M in 1948.

A milestone in the development of range science as both a discipline and as a profession was the formal organization of the American Society of Range Management at its first annual meeting in Salt Lake City in January 1948, at which time it had 500 members. Undergoing a name change in 1971, the Society for Range Management now has grown to a membership of about 5,500. Through its annual meetings, its position statements, its regional section activities, and its Journal of Range Management, Rangelands, and other serial publications, the society has refined and promoted the principles of range science.

Chapter 2

RANGE SCIENCE LITERATURE SEARCHING

LITERATURE SEARCHING PROCEDURES

Although an intensive literature search may not always be required to meet the need for information, this section presumes that such a search will be required and emphasizes how to proceed on range science topics. Literature search, by definition, means a systematic and exhaustive search for published materials bearing on a specific subject, including the preparation of abstracts of pertinent data for subsequent use. The need giving rise to the literature search may somewhat modify procedures to be followed in literature searching, but they should always be methodical and precise.

The following steps are suggested for making intensive searches in range science literature. All literature searches must generally proceed through number six; some will require proceeding through all steps.

1. Select and refine the search topic. The topic will generally be dictated by the need and ultimate use of the results of the literature search, but considerable latitude may be provided. All topics should be refined before searching begins and often during the search as well. Searches should be confined to limited topics; unrefined, broad topics become unwieldy and burdensome by requiring the handling of unnecessarily large volumes of literature. Yet, the topic should not be so confining that no latitude is permitted.

2. Provide background for the search topic. Insure that terms within and closely allied to the selected topic have been properly defined and understood. For assistance in defining range science terminology, refer to the list of "Glossaries of Range Science and Related Fields." Developing a list of synonyms and related key words will also materially expedite later stages of the search.[12] A major part of this step will consist of reading about the topic in range science textbooks and other reference materials. Refer to list of "Key Range Science Reference Literature." Note names of scientists, institutions, and organizations associated with the topic for further reference.

3. Search the library card catalog. The library card catalog should be the initial tool used to locate range science literature in libraries. Most card catalogs include both author and subject entries and sometimes title entries. However, it is important to note that the card catalog includes entries for mono-

graphs and selected other forms of literature, but will not list articles within periodicals nor generally individual titles in monographic series, proceedings, and reviews. The card catalog will seldom provide over 15 percent of key references on range science topics.

4. Browsing in library stacks. Library call numbers taken from the card catalog or selected on the basis of an understanding of the library classification system for literature used in the library provide the principal means of locating appropriate sections in open stacks for browse searching. (Explanations of the three principal filing systems used in American research libraries are provided in a subsequent section.) Although incomplete and ineffective by itself as a literature search technique, browsing can be additive when used with other techniques. The principal problems in placing reliance on browsing include (a) no classification system adequately treats the expanded field of range science, (b) many valuable literature items are filed in other places, i.e., periodicals, serials, etc., and (c) only books not being presently used--often the least helpful--are in the stacks. The last problem can be circumvented by using the shelf list, a card file of all books arranged by call number in the same order as the books on the shelf.

5. Check printed range science bibliographies. Bibliographies ranging from those compiled on narrow topics to those covering the entire field of range management are available in printed form. A selected list of published bibliographies has been included in a later section entitled "Published Bibliographies for Range Science"; also included are two subject bibliographies: "Subject Bibliography of North American Range Science" and "Introductory Bibliography of Foreign Range Management." Two published bibliographies that attempted to comprehensively cover the entire field of range science are: (1) A SELECT BIBLIOGRAPHY ON MANAGEMENT OF WESTERN RANGES, LIVESTOCK, AND WILDLIFE issued in 1938 and including 8274 entries and (2) U.S.-CANADIAN RANGE MANAGEMENT, 1935-1977: A SELECTED BIBLIOGRAPHY ON RANGES, PASTURES, WILDLIFE, LIVESTOCK, AND RANCHING published in 1978 and including about 20,600 entries.

Although very helpful to the literature searcher, such bibliographies can never be considered complete, particularly in areas peripheral to the core areas of range science. Another useful source of compiled bibliography is the literature cited sections or bibliography sections found in review articles, textbooks, and most other technical literature. However, even these should never be considered as exhausting the literature. And they, of course, can never be more up-to-date than the date of publication of the literature in which they are found.

6. Utilize indexing and abstracting literature. Comprehensive literature searches will eventually require the use of literature that variously abstract, index, list or catalog scientific literature as released. Intensive use of these aids will invariably locate many pertinent items not found previously. Although time-consuming and requiring considerable skill on the searcher's part, literature searching must at least progress through this step before being considered reasonably complete. A listing, evaluation, and suggestions for using those considered most useful in searching range science topics are given in the section on "Indexing and Abstracting Literature." However, because of the time lag between date of publication of the original literature and their appearance in such finding aids--even though reduced considerably in recent years--resort

should also be made to using appropriate "Current Contents" and manually searching recent issues of subject matter journals and monographic serials known to emphasize topics pertinent to the literature search being made.

7. Utilize computer retrieval services. Computer-based information systems may be used to complement or partially replace the manual searching of abstracting and indexing journals. Mechanization has greatly expedited the arranging, sorting, cumulating, and printing out of scientific literature bibliography. References so arranged and stored on computer disks or tapes are readily converted into bibliographies based on subject, author, or issuing agency. Such systems can be used to provide custom searches on specific subjects as requested or to assemble bibliographies on selected topics for publication. Some systems provide only bibliographic information on each reference, while others can provide source data or abstracts as well. Computerized information retrieval systems generally include only recent literature, i.e., CAIN (Cataloging and Indexing System) references are mostly since 1970. Both computer systems and bibliographic journals are subject to the serious limitations of key words in literature titles not representing the actual material covered.

8. Utilize citation indexes. This unique approach to literature searching is based on the assumption that articles cited in a publication reflect the content of the publication and are useful in indexing it. Each article listed in a citation index is provided with (1) a list of cited references, i.e., older articles cited in the article in question, and (2) a list of citing references, i.e., more recent articles that have cited the article in question. A single article on the subject then serves as a starting point in locating others.

9. Correspond with known experts. Individual scientists and research organizations involved in in-depth studies on the topic being searched may be able to point to additional literature on the topic, provide consultation on the search topic, or may even be willing to share unpublished data. However, this approach must be used with good judgment and willingness to respect the confidentiality of unpublished data unless expressly released for public use.

Special Aids on Literature Searching

Several published works, including both general references and information source books related to range science, provide general background as well as specific information on literature searching. Those in the following list are recommended.

1. Bottle, R.T., and H.V. Wyatt (Eds.). 1971 (2nd Ed.). THE USE OF BIOLOGICAL LITERATURE. Butterworths, London. 379 p.

2. Bowker (R.R.) Company, Serials Bibliography Dept. 1978 (5th Ed.). IRREGULAR SERIALS AND ANNUALS: AN INTERNATIONAL DIRECTORY. R.R. Bowker Co., New York. 1,396 p.

   Includes serials and continuations such as proceedings, transactions, advances, progresses, reports, yearbooks, handbooks, annual reviews, and monographic series, which constitute the "twilight" area between books and periodicals.

3. Bowker (R.R.) Company, Serials Bibliography Dept. 1979 (18th Ed.). ULRICH'S INTERNATIONAL PERIODICALS DIRECTORY. R.R. Bowker Co., New York. 2,156 p.

4. Duke, Dorothy Mary. 1962. AGRICULTURAL PERIODICALS PUBLISHED IN CANADA, 1836-1960. Information Div., Can. Dept. Agric., Ottawa, Ont. 101 p.

5. ENCYCLOPEDIA OF ASSOCIATIONS. Vol. 1: NATIONAL ORGANIZATIONS OF THE U.S. 1980 (14th Ed.). Gale Research Co., Detroit, Mich. 1,566 p.

    Includes mostly U.S., nonprofit membership organizations.

6. Greenberg, Howard (Managing Ed.). 1978 (6th Ed.). THE STANDARD PERIODICAL DIRECTORY. Oxbridge Communications, Inc., New York. 1,680 p.

    Includes regular serials published at least biennially.

7. Grogan, Denis. 1976 (3d Ed., Rev.). SCIENCE AND TECHNOLOGY: AN INTRODUCTION TO THE LITERATURE. Linnet Books, Hamden, Conn. 343 p.

8. Malinowsky, H. Robert, Richard A. Gray, and Dorothy A. Gray. 1976 (2nd Ed.). SCIENCE AND ENGINEERING LITERATURE: A GUIDE TO REFERENCE SOURCES. Libraries Unlimited, Inc., Littleton, Colo. 368 p.

9. Morris, Jacquelyn M., and Elizabeth A. Elkins. 1978. LIBRARY SEARCHING: RESOURCES AND STRATEGIES. Jeffrey Norton Pub., New York. 129 p.

10. National Academy of Sciences. 1971 (9th Ed.). SCIENTIFIC, TECHNICAL, AND RELATED SOCIETIES OF THE UNITED STATES. National Academy of Sciences, Washington, D.C.

11. Sheehy, Eugene P., Rita G. Keckeissen, and Eileen McIlvaine (Comp.). 1976. (9th Ed.). GUIDE TO REFERENCE BOOKS. American Library Assoc., Chicago. 1015 p.

    An expansion and updating of the 8th edition of Winchell's GUIDE TO REFERENCE BOOKS; includes an agriculture section and a biological science section.

12. Smith, Roger C., and Reginald H. Painter. 1966 (7th Ed.). GUIDE TO THE LITERATURE OF THE ZOOLOGICAL SCIENCES. Burgess Pub. Co., Minneapolis, Minn. 238 p.

13. Smith, Roger C., and W. Malcolm Reid. 1972 (8th Ed.). GUIDE TO THE LITERATURE OF THE LIFE SCIENCES. Burgess Pub. Co., Minneapolis, Minn. 166 p.

14. U.S. Library of Congress, National Referral Center. 1972. A DIRECTORY OF INFORMATION RESOURCES IN THE UNITED STATES: BIOLOGICAL SCIENCES. U.S. Govt. Print. Office, Washington, D.C. 577 p.

15. Walford, A.J. (Ed.). 1973 (3rd Ed.). GUIDE TO REFERENCE MATERIALS. Vol. I: SCIENCE AND TECHNOLOGY. The Library Assoc., London. 615 p.

Includes sections on biological sciences, general botany, zoology, and agriculture.

LIBRARY CLASSIFICATION SYSTEMS

Printed literature in libraries is classified by a combination of letters and numbers into groupings of common characteristics, for example, subject, author, or publisher (issuing organization) or a combination of these, to provide a definite placement in the library and enable rapid retrieval by library users. The principal library classification systems used in the United States are the Dewey Decimal, Library of Congress, and Superintendent of Documents systems. A brief description of each system as it pertains to range science literature searching follows.

Dewey Decimal Classification System

The Dewey system, first published in 1876, is the most widely used and best known library classification system in the United States. This system utilizes ten main classes, including one section for general works and nine sections representing fields of human knowledge, and a hierarchical arrangement of subdivisions. The decimal numbering system provides for a minimum, three-digit subject classification, followed by a decimal point and up to 22 additional digits for additional subject matter subdivision. This system has the advantage of being described and indexed in a three-volume work, DEWEY DECIMAL CLASSIFICATION AND RELATIVE INDEX.[13] An adaptation of the Dewey Decimal System, referred to as the Universal Decimal Classification System, has become popular in many European countries.

The major disadvantage of the Dewey Decimal System is that it has not been updated sufficiently to provide satisfactory categories for range science literature. As a result, the literature of range science is widely spread through the 600s, 500s, and 300s, thereby providing only minimum opportunities for literature browsing. The application of the Dewey system to range science literature is further explained in outline form.

10 MAIN CLASSES

000 Generalities
100 Philosophy and Related Disciplines
200 Religion
300 Social Sciences
400 Language
500 Pure Sciences
600 Technology (Applied Sciences)
700 The Arts
800 Literature
900 General Geography and History

DIVISIONS OF PRIMARY APPLICATION TO RANGE SCIENCE

- 330 Economics
  - 333 Land economics
    - 333.1 Public control of land
    - 333.3 Individual control of land
    - 333.7 Land utilization
      - 333.72 Conservation (of natural resources)
      - 333.74 Pasture (grazing) lands
      - 333.75 Forest lands
      - 333.76 Agricultural and other rural lands
      - 333.78 Recreational lands
    - 333.9 Utilization of other natural resources (including water)
  - 338 Production
    - 338.1 Agriculture (ranch economics)
- 550 Earth Sciences
  - 551.4 Geomorphology (including physical geography and hydrology)
  - 551.5 Meterology
  - 551.6 Climatology and weather
- 570 Anthropology and biological sciences
  - 574 Biology (including botany, zoology, and natural history)
    - 574.1 Physiology (biophysiological)
    - 574.5 Ecology (bioecology, ecosystems)
  - 579 Collection and preservation of biological specimens
- 580 Botanical Sciences
  - 581 Botany
    - 581.1 Physiology of plants
    - 581.2 Pathology of plants
    - 581.5 Ecology of plants
    - 581.6 Economic botany
      - 581.69 Poisonous plants
    - 581.9 Geographic treatment (including regional floras)
  - 582 Spermatophyta
  - 583-4 Angiospermae
  - 585 Gymnospermae
  - 586-9 Other plant divisions
- 590 Zoological Sciences
  - 591 Zoology
    - 591.1 Physiology of animals
    - 591.5 Ecology of animals
    - 591.9 Geographic treatment

599 Mammalia
599.323 Rodentia
599.735 Ruminantia
630 Agriculture and Related Technologies (630.5-7 includes general agriculture serials)
631 General Agriculture
631.2 Agricultural structures (631.27 includes fences)
631.3 Agricultural tools, machinery, and equipment
631.4 Soil and soil conservation
631.5 Cultivation and harvesting (including plant breeding and propagation)
631.8 Fertilizers and soil conditioners
632 Plant injuries, diseases, pests (including plant control)
633 Field crops
633.2 Forage crops (emphasizing grasses)
633.3 Legumes and other forage crops
634 Orchards, fruits, forestry
634.9 Forestry
634.99 Other aspects; including grazing and pasture use
636 Animal husbandry (636.081-9 includes general care, selection, nutrition, feeds, and veterinary medicine)
636.1 Equines; horses
636.2 Ruminants; bovines; cattle (includes bison, antelope, and other Bovoidea; deer, elk, moose, caribou, and other Cervoidea)
636.3 Smaller ruminants, sheep (includes goats)
639 Nondomesticated animals (wildlife; game management)
639.1 Hunting and trapping
639.9 Conservation of biological resources

Library of Congress Classification System

This classification system, begun in 1900, was tailored to the literature collection in the Library of Congress. The main subject classes are designated by a single capital letter, with a second capital letter being used to designate the secondary divisions of each main class. Further subject matter subdivision is by arabic numbers, but an alphabetical listing commonly replaces the more useful hierarchical subject matter arrangement.

Many research libraries have recently shifted from the Dewey Decimal to the Library of Congress (LC) system. Since the large Ohio College Library Center data base is coded with the LC system, libraries with LC classification can utilize this system for cataloging, reference services, and production of cards for the card catalog. This system also has unlimited expansion features, and new subjects can be inserted readily.

The LC system is described in 32 separate classification schedules, mostly main class summaries. However, the lack of a combined index to all schedules is a major shortcoming of the system. It resembles the Dewey Decimal System in that adequate categories have not been provided for range science, resulting in its pertinent literature also being widely dispersed in the LC system, particularly in the Q (Science) and S (Agriculture) main classes. The application of the LC system to range science literature is explained in the following outline:[14]

MAIN CLASSES (Summary)

A General Works
B Philosophy, Psychology, Religion
C Auxillary Sciences of History
D-F History
G Geography, Anthropology, Recreation
H Social Sciences
J Political Science
K Law
L Education
M Music
N Fine Arts
P Language and Literature
Q Science
R Medicine
S Agriculture
T Technology
U-V Military and Naval Sciences
Z Bibliography, Library Science

DIVISIONS OF PRIMARY APPLICATION TO RANGE SCIENCE

GB Physical geography
- 651-2998 Hydrology, water
  - 906 Watersheds

GV Recreation
- 193-200.5 Outdoor recreation (also 182.2)

HD Land, agriculture, industry
- 101-1395 Land (including real estate and land tenure)
  - 216-243 Public lands (including grazing)
- 1405-2206 Agriculture (including agricultural economics, water rights)
  - 1635-1641 Pasture lands

QH Natural history; landscape protection (including wildlife conservation)
- 540-549 General ecology

QK Botany
- 75-77 Herbariums
- 91-97 Classification
- 100 Poisonous plants
- 101-474.5 Phytogeography (including floras)
- 475-497 Spermatophyta
- 641-673 Plant anatomy
- 710-899 Plant physiology
- 901-938 Plant ecology

QL Zoology
- 81.5-84.77 Wildlife conservation

	101-345	Geographical distribution
	351-355	Classification and nomenclature
	671-699	Birds
	700-739.2	Mammals
	750-991	Animal behavior, morphology, anatomy, and embryology
S	Agriculture General	
	560-575	Farm management, farm economics
	591-599	Soils
	622-627	Soil conservation
	631-667	Fertilizers and soil improvement
	671-760	Farm machinery and engineering
	900-972	Conservation of natural resources
SB	Plant culture	
	114-117	Seeds
	119-124	Propagation (including breeding)
	193-207	Forage crops (including ranges and range watersheds, pastures, harvested roughage)
	481-485	Parks and public reservations
	611-615	Weeds
	617-618	Poisonous plants
	621-795	Plant pathology
	993-999	Noxious and useful animals
SD	Forestry	
	425	Floods, forests, and water supply (including forest watersheds)
	426-428	Forest reserves (including grazing)
SF	Animal culture	
	85	Stock ranges
	95-99	Feeds and feeding, animal nutrition
	105-109	Breeds and breeding
	191-219	Beef cattle
	277-318	Horses
	371-379	Sheep, wool
	381-385	Goats
	399-401	Semidomesticated (including deer, muskox, reindeer)
	508-510	Game bird culture
	600-988	Veterinary medicine
SK	Hunting	
	295-305	Big game hunting
	311-335	Bird hunting
	351-579	Wildlife management, game protection
	601-605	Dude ranching

Superintendent of Documents Classification System[14]

This system has been used since 1903 to classify the public documents issued by the various branches of the U.S. federal government and placed in the Library of the Public Documents Department. In this system, literature is filed not on the basis of subject matter but by issuing divisions of the federal government. Each executive department and agency, the judiciary, Congress, and the major independent establishments are assigned a place in the classification system, with subordinate units being grouped under the parent organization.

The Superintendent of Documents system uses a combination of letters and numbers to designate each document. Each notation includes three segments: (1) the issuing organization, (2) the category designation (placed between a period and a colon), and (3) the book number (placed after the colon). In the first segment the parent organization is designated by one or two letters, followed by one or two arabic numbers to designate the subordinate unit.[15] For example, the notation for USDA, Forest Service, Forest Resource Report 19, "The Nation's Range Resources," is A 13.50:19. "A 13" indicates that USDA's Forest Service is the issuing organization; "50" indicates the Forest Resource Report Series; and "19" indicates issue number 19 in the series.

The major disadvantage of the Superintendent of Documents system is that it does not bring together information of similar subject matter, except to the extent of common interests within a particular branch of federal government. Thus, browsing in the documents section of the library is mostly unfruitful, and call numbers (notations) must be obtained from the monthly catalog[16] that serves as an index to federal documents. Another disadvantage is that changes in the organizational structure of federal government also requires a corresponding change in the classification system. An advantage of the system is that all federal documents are classified and notated prior to publication; this is of distinct advantage to libraries designated as federal documents depositories.

The principal application of the Superintendent of Documents system to range science is shown in the following selected divisions assigned to the U.S. Departments of Agriculture and Interior:

DIVISIONS OF PRIMARY APPLICATION TO RANGE SCIENCE

A U.S. Department of Agriculture

(Prior January 1978)

A1	Secretary's Office
A13	Forest Service
A17	National Agricultural Library
A43	Federal Extension Service
A57	Soil Conservation Service
A77	Agricultural Research Service
A93	Economic Research Service

(After January 1978)

A1	Secretary's Office
A13	Forest Service
A106.1	SEA-Technical Information Systems
A106	SEA-Extension
A57	Soil Conservation Service
A106	SEA-Agricultural Research
A105	Economics, Statistical, and Cooperative Service

I. U.S. Department of Interior

11	Secretary's Office
120	Bureau of Indian Affairs
122	Departmental Library
127	Bureau of Reclamation
129	National Park Service
149	Fish and Wildlife Service
153	Bureau of Land Management
166	Bureau of Outdoor Recreation

EP Environmental Protection Agency

INDEXING AND ABSTRACTING LITERATURE

Indexes are the keys to the contents of scientific literature. Cumulative indexes to individual journals are helpful but have been prepared only for a few journals, and such indexes may be incomplete. Cumulative indexes to monographic and other serial literature are even less common. However, indexes to a single serial can be expected to cover only a small portion of the literature in a given field. Needed for effective literature searching are indexes that index the majority of literature in the field and that cover many sources.

Indexes are required only to provide sufficient information about each literature item that it can be identified and traced. Indexes depend on the use of subject headings and the effectiveness of the titles and their keys words. Such indexes can only be really effective if the titles adequately represent the contents of the literature; enriching titles with additional key words and phrases can increase indexing effectiveness somewhat. The indexing of scientific literature would be enhanced materially if authors would consider the problems of indexing before composing titles to articles.

Literature abstracts allow the contents of scientific literature to be explained more fully than literature indexes. Indicative abstracts provide additional explanation as to the contents of literature; informative abstracts actually summarize the principle data and conclusions in the original publication. Abstracts aid literature searchers in deciding whether the original literature items should be located and used.

No indexing or abstracting journal can be considered complete on any given subject. Also, much depends on the kind and effectiveness of the indexes used in the indexing and abstracting literature. Thus, more than one indexing or abstracting source should be used, and this should be combined with other approaches to locating literature for acceptable completeness.

Few indexes will present over 50 percent of the literature on a subject for the time period covered, and great skill and diligence of the literature searcher will be required for 50 percent retrieval. Retrieval of 75 percent of the literature on a subject from using a single or even combination of indexing and abstracting literature is probably unusual.

Bibliography of Agriculture is apt to be the first and foremost indexing journal used in exhaustive literature searches on range science subjects, particularly for subjects on applied aspects of range management.[17] It has the advantage of representing the total incoming literature to the SEA-Technical Information Systems (U.S. National Agricultural Library). However, complete coverage of most subjects, particularly range science subjects only indirectly related to production agriculture, will often require use of additional abstracting and indexing literature described below.

ABSTRACTS OF RECENT PUBLISHED MATERIAL ON SOIL AND WATER CONSERVATION
Publisher: Agricultural Research Service, USDA, Washington, D.C. 20250
Publication: began 1955 with issue No. 1; last issue, No. 42 in 1967; published irregularly.
Coverage and format: selected literature on soil and water conservation; citations along with abstracts entered under primary and secondary subject matter headings; table of contents provided in front of each issue.
Indexes: an author index is included beginning with issue No. 32.

AMERICAN BIBLIOGRAPHY OF AGRICULTURAL ECONOMICS
Publisher: American Agricultural Economics Association, c/o Department of Agricultural Economics, University of Kentucky, Lexington, Ky. 50506; prepared by the American Agricultural Economics Documentation Center, USDA, Washington, D.C. 20250.
Publication: began in 1971; frequency, 4-6 issues per year; discontinued with completion of Volume 4 in 1974 but information subsequently included in Bibliography of Agriculture.
Coverage and format: U.S. and Canadian literature on agricultural economics including monographs, serial publications, journal articles, mimeographed reports, and papers of conferences and symposia; citations and abstracts arranged alphabetically by author under eight subject matter headings.
Indexes: subject index and author index in each issue.

BIBLIOGRAPHY OF AGRICULTURE
Publisher: (1) U.S. National Agricultural Library (NAL), USDA, Washington, D.C. 20250 (through 1969); (2) CCM Information Corporation and Macmillan Information, 866 Third Avenue, New York City, N.Y. 10222 (1970-1974); and (3) The Oryx Press, 2214 North Central Ave., Phoenix, Ariz. 85004 (1975 to present).
Publication: began 1942 with Vol. I; present volume, Vol. 42 (1978); published monthly, 1970 to present; December issue devoted to volume indexes prior to 1970.
Coverage and format: comprehensive coverage of worldwide literature on agriculture and allied sciences; monthly issues based on indexing records of acquisitions at the National Agricultural Library (name changed in 1978 to SEA-Technical Information Systems) prepared for its CAIN computer system; includes jour-

nal articles, pamphlets, special reports, government documents, proceedings, etc.; generally excludes editorials, letters to editors, and general columns; English translations provided for foreign language titles; items listed may be purchased from NAL in microfilm or photoprint; in the main entry section of each monthly issue, each bibliographic entry is assigned to one or more broad subject matter categories (76 categories in 1977) and includes a complete bibliographic citation and a six-digit identification number (from 1970 through 1975 subject headings in the main entry section varied from 17 to 66; prior to 1970 ten primary and a number of secondary and tertiary subject headings were used); subject matter headings are shown in table of contents; main entry subsections (omitted 1970-1975 inclusive) list respectively USDA publications, state agricultural experiment station publications, state agricultural extension service publications, FAO publications, and translated publications (identification numbers dropped as complete bibliographic citation given and each entry also entered in main entry section).
Indexes: each monthly issue now provided with a geographic index, a corporate author index, a personal author index, and a subject index; annual cumulation is a subject and author index to entries in the 12 monthly issues (11 issues prior to 1970) with entries referenced by identification numbers to the monthly issues; in the monthly and cumulation subject indexes all titles are complete, entries are based on single title words or title enrichment words printed in italics, secondary words either preceding or following the title words and printed in bold face; a Bibliography of Agriculture Retrospective Cumulation on Microfiche, 1970-1977 (Volumes 34-41) followed by annual updates on microfiche began publication in the fall, 1978.

BIOLOGICAL ABSTRACTS
Publisher: BioSciences Information Service of Biological Abstracts, 2100 Arch St., Philadelphia, Pa. 19103.
Publication: began 1927; present volumes (1978), Vol. 65 and 66; semimonthly (beginning 1960), monthly through 1959.
Coverage and format: worldwide coverage of the biological sciences; includes both serial and monographic literature; citations and accompanying abstracts placed under major and minor subject headings; subject guide in each issue.
Indexes: indexes in each issue include author index, biosystematic index (gross taxonomic categories), generic index (Latin names), concept index, and subject index; corresponding indexes cumulated on semiannual and annual basis.

BIOLOGICAL AND AGRICULTURAL INDEX (formerly Agricultural Index to August 1964, Vol. 1-18)
Publisher: H.W. Wilson Co., Bronx, N.Y. 10452.
Publication: began 1919 with coverage back to 1916; present volume, Vol. 32 (August 1977-July 1978); published monthly except August, cumulated quarterly and biannually (1916-1963) or annually (1964 to present).
Coverage and format: selected English-language literature; has included only periodicals in biological, agricultural, and related sciences beginning in August 1964; prior to August 1964 also included state agricultural experiment station and USDA publications, some state agricultural extension publications, and some other miscellaneous literature.
Indexes: no indexes; titles listed under numerous alphabetical subject headings; no author entries; easy to use because of subject matter arrangement.

BIORESEARCH INDEX (formerly BioResearch Titles, 1965-1966)
Publisher: BioSciences Information Service of Biological Abstracts, 2100 Arch St., Philadelphia, Pa. 19103.
Publication: began 1965; present volume, Vol. 15 (1978); issued monthly.
Coverage and format: extends coverage but does not overlap with BIOLOGICAL ABSTRACTS; covers symposia, book chapters, scientific meetings, and newer and lesser-known journals; entries grouped by source, i.e., journal, other serial, etc.; bibliographic information but not titles are provided for each entry.
Indexes: author, biosystematic, generic, concept, and subject indexes included in each issue; indexes cumulated in annual volumes also.

CANADIAN GOVERNMENT PUBLICATIONS MONTHLY CATALOG
Publisher: Supply and Services Canada, Publications Division, Ottawa, Ont. K1A 0S9.
Publication: began 1953; present volume, Vol. 26 (1978); monthly and annual catalogs.
Coverage and format: includes all Canadian government publications, documents, and papers not confidential; includes publications of Agriculture Canada and Fisheries and Environment Canada (including Canadian Wildlife Service and Canadian Forestry Service).
Indexes: a general index and an index to periodicals in each monthly issue; annual catalog includes cumulated general index and index to periodicals.

DICTIONARY CATALOG OF THE NATIONAL AGRICULTURAL LIBRARY (now SEA-TIS)
Publisher: Rowman and Littlefield, 81 Adams Drive, Totowa, N.J. 07512; prepared by the U.S. National Agricultural Library.
Publication: first 73 volumes covering 1862-1965 period published 1967-1970; a 12-volume supplement entitled NATIONAL AGRICULTURAL LIBRARY CATALOG, 1966-1970, also published by Rowman and Littlefield, Totowa, N.J.
Coverage and format: first 72 volumes include over 1.7 million cards and comprise an A to Z sequence of authors, titles, and subjects of monographs, serials, and analyticals as cataloged at the National Agriculture Library; Vol. 73 comprised of translations of periodical articles; updated by monthly supplements beginning in 1966 entitled "National Agricultural Library Catalog"; use of monthly supplements made mostly unnecessary since materials included in BIBLIOGRAPHY OF AGRICULTURE.
Indexes: no indexes; supplanted by alphabetical arrangement of entry cards.

FERTILIZER ABSTRACTS
Publisher: National Fertilizer Development Center, Tennessee Valley Authority, Muscle Shoals, Ala. 35660.
Publication: began 1968; present volume (1978), Vol. 11; issued monthly.
Coverage and format: includes worldwide literature on fertilizer technology, marketing, and use from about 400 journals, bulletins, and technical reports; bibliographic citation and abstract provided for each entry.
Indexes: author and subject indexes in each issue; author and subject indexes also cumulated annually.

HERBAGE ABSTRACTS
Publisher: Commonwealth Agricultural Bureau, Farnham Royal, Slough SL2 3BN, UK; prepared by the Commonwealth Bureau of Pastures and Field Crops, Hurley, Maidenhead, Berks SL6 5LR, UK.
Publication: began 1931; present volume (1978), Vol. 49; issued monthly (1973 to present), quarterly (1950-1972), and bimonthly (prior to 1950).
Coverage and format: worldwide literature on grassland husbandry and fodder crop production (pastures, rangelands, herbage plants); citations and accompanying indicative or informative abstracts arranged under primary and secondary subject headings; subject headings given in table of contents; includes book reviews; many issues contain a review article.
Indexes: subject index and author index now included in each issue; annual subject index and author index in all volumes.

INDEX TO SCIENTIFIC REVIEWS
Publisher: Institute for Scientific Information, 325 Chestnut St., Philadelphia, Pa. 19106.
Publication: began 1975; issued semiannually and in permanent annual volumes.
Coverage and format: worldwide coverage; review articles found in review journals, selected articles from other journals, and monographic review series; taken from the source index of Science Citation Index where designated by R."
Indexes: included in each semiannual and annual volume are citation indexes (arranged alphabetically by cited author), corporate index (arranged alphabetically by affiliated organization), source index (arranged alphabetically by citing author), and permuterm subject index (a coterm, title-word index).

MONTHLY CATALOG OF U.S. GOVERNMENT PUBLICATIONS
Publisher: U.S. Superintendent of Documents, Washington, D.C. 20402.
Publication: began in 1900; issued monthly.
Coverage and format: includes all U.S. Government publications except administrative and confidential papers; entries listed by originating agencies by Superintendent of Documents classification system; serves as index to materials cataloged under this system.
Indexes: general alphabetical index in each monthly issue through 1973 with subject (derived from LIBRARY OF CONGRESS SUBJECT HEADINGS), title, and author indexes beginning in 1974; annual cumulative subject matter index through 1973 with subject, author, and title indexes beginning in 1974; semiannual subject, author, and title indexes begun in 1976.

MONTHLY CHECKLIST OF STATE PUBLICATIONS
Publisher: Library of Congress, Washington, D.C. (sold by U.S. Superintendent of Documents).
Publication: began 1910; present volume (1978), Vol. 69; issued monthly.
Coverage and format: state documents including monographs, monographs in series, and annual reports received by the Library of Congress; periodicals are listed semiannually in the June and December issues; titles arranged alphabetically by state and then issuing department.
Indexes: annual subject index to the monographic literature.

NUTRITION ABSTRACTS AND REVIEWS
Publisher: Commonwealth Agricultural Bureau, Farnham Royal, Slough SL2 3BN, UK; prepared by Commonwealth Bureau of Nutrition, Rowett Research Institute, Buckburn, Aberdeen, AB2 9SB, Scotland.
Publication: began 1931; present volume (1978), Vol. 49; issued monthly (1973 to present), quarterly (1972 and previously).
Coverage and format: worldwide literature including chemical composition of feeds, physiology of nutrition, feeding of animals, etc.; citations and accompanying abstracts arranged under subject matter headings; original review articles, book reviews, and reports included in many issues; publication divided into two series beginning in 1977: A. Human and Experimental and B. Livestock Feeds and Feeding.
Indexes: cumulative annual author index and subject index included in all volumes; author index and subject index now included in each issue.

SCIENCE CITATION INDEX
Publisher: Institute for Scientific Information, 325 Chestnut St., Philadelphia, Pa. 19106.
Publication: began 1963 with annual volumes back to 1961; present volume, 1978 Annual Volume; issued quarterly with annual cumulations; quinquennial cumulations also for 1965-1969 and 1970-1974.
Coverage and format: covers source journals (2717 in 1976) in science, medicine, agriculture, and technical and behavioral sciences; nonjournal materials such as symposia, monographic series, and multiauthored, themic books added beginning in 1977; should be used as a supplement but not as a replacement for conventional indexes.
Indexes: quarterly issues and annual and quinquennial cumulations utilize the following indexes (sections): (1) Source Index - a list of citing literature for the period covered arranged by primary (sole or first) authors with secondary authors being cross referenced; original items can then be checked to see what literature was cited; (2) Citation Index - an index of literature references cited during the period covered arranged by cited author; shows who was citing it and in what publications it was cited; (3) Permuterm Subject Index - a permuted title-word index in which every significant word is paired with every other significant word in the same title; name of author provided, thus allowing each article to be completely identified in the Source Index section.

SELECTED WATER RESOURCES ABSTRACTS
Publisher: Water Resources Scientific Information Center, Office of Water Research and Technology, USDI, Washington, D.C. 20240.
Publication: began 1968; present volume (1978), Vol. 11; issued semimonthly.
Coverage and format: worldwide literature on characteristics, conservation, control, use, and management of water; literature includes monographs, journal articles, reports, and other publication forms; citations and abstracts arranged by subject fields and groups.
Indexes: subject, author and organizational indexes included in each issue; indexes cumulated annually.

SOILS AND FERTILIZERS (ABSTRACTS)
Publisher: Commonwealth Agricultural Bureau, Farnham Royal, Slough SL2 3BN,

UK; prepared by Commonwealth Bureau of Soils, Rothamsted Experiment Station, Harpenden, Herts, AL5 2JQ, UK.
Publication: began 1938; present volume (1978), Vol. 41; issued monthly (beginning 1973), quarterly (prior to 1973).
Coverage and format: world literature of pedology, soil physics, classification, biology, grasslands, etc.; citations and abstracts arranged by primary and secondary subject headings; includes book reviews and review articles.
Indexes: author index and subject index now in each issue; annual cumulative author index and subject index in each volume.

WEED ABSTRACTS
Publisher: Commonwealth Agricultural Bureau, Farnham Royal, Slough SL2 3BN, UK; prepared by Weed Research Organization, Yarnton, Oxford, OX5 1PF, UK.
Publication: began 1952; present volume (1978), Vol. 27; issued monthly (beginning 1973), bimonthly (prior to 1973).
Coverage and format: world literature on weeds, weed control, and allied subjects; citations and informative or indicative abstracts arranged by primary and secondary subject headings; book reviews included; some issues contain original review articles.
Indexes: annual species, subject, and author indexes; species, subject, and author indexes now included in each issue.

WILDLIFE REVIEW
Publisher: USDI, Fish and Wildlife Service, Colorado State University, Fort Collins, Colo., 80523 (previously Patuxent Wildlife Research Center, Washington, D.C. 20420).
Publication: began 1935; issues not combined into volumes; Issue No. 167 issued December 1977; irregular but began being issued quarterly since about 1960.
Coverage and format: worldwide literature on wildlife biology and management; entries including citation and abstract (if included) arranged by primary and secondary subject headings.
Indexes: an author index and a geographic index now included in each issue; subject entry made only through table of contents; book reviews included. WILDLIFE ABSTRACTS: a cumulative bibliography and index (abstracts excluded) to entries in WILDLIFE REVIEW is entitled WILDLIFE ABSTRACTS, published by the U.S. Fish and Wildlife Service; issues have been published for 1935-51, 1952-55, 1955-60, and 1961-70; entries are arranged under primary and secondary categories; author and subject indexes are included.

WORLD AGRICULTURAL ECONOMICS AND RURAL SOCIOLOGY ABSTRACTS
Publisher: Commonwealth Agricultural Bureau, Farnham Royal, Slough SL2 3BN, UK; prepared by Commonwealth Bureau of Agricultural Economics, Dartington House, Little Clarendon St., Oxford, OX1 2HH, UK.
Publication: began 1959; present volume (1978), Vol. 20; monthly (beginning 1973), quarterly (prior to 1973).
Coverage and format: worldwide literature of agricultural economics and rural sociology; entries including citations and abstracts arranged by primary and secondary headings; primary headings pertinent to range science include economics of production and finance and credit; book reviews included.
Indexes: annual author and subject indexes for each volume; author and subject indexes in each issue.

COMPUTERIZED LITERATURE RETRIEVAL

The information explosion is occurring at an ever increasing rate. This is no less true for the field of range management than it is for other scientific disciplines. The information, however, for a developing discipline such as range science is spread throughout various other disciplines on which the science is based. This makes information gathering a difficult task since numerous sources must be located, inventoried, and reviewed. Fortunately, along with the explosion of information is an associated dramatic increase in technology of high-speed computers. Such computers are being marshalled into use for handling large amounts of information in an easily retrievable format.

Numerous data bases exist covering all phases of food and agriculture. Information on range science exists under a wide variety of categories including animal science, chemistry, economics, sociology, entomology, environmental sciences, food and human nutrition, forestry, physical sciences, plant and soil sciences, watershed management, as well as the general research and management fields. Computerized data bases have been developed by individual researchers and scientists, federal agencies, universities, and state experiment stations. Private industry, also, is heavily involved in the development of computerized literature retrieval systems.

Nineteen agencies within the Department of Agriculture maintain information on automated on-line data systems.[18] This is a part of the department's activities in meeting responsibilities to provide other governmental agencies, the private sector, and the public with current and reliable agricultural data. Many of these are for in-house management of information and are not readily accessible to the general public.

Types of Data Bases

There are two major types of information systems, citation data bases and information data bases. Citation data bases are generally literature-oriented and are collections of bibliographic material describing books, journal articles, and technical reports: author, title, year of publication, index terms, and sometimes an abstract. A prominent example of a citation data base is the Cataloguing and Indexing System (CAIN) developed by the National Agricultural Library (now SEA-Technical Information Systems). Biological Abstracts, Chemical Abstracts, and Engineering Index are other important examples.

The information data base is a collection of identifiable facts, either unevaluated or evaluated, that are filed in a logical grouping. Evaluated data bases may be discipline-oriented or activity-oriented. An example of this type of data base would be a land inventory by soil classes such as that developed by the Soil Survey Division of the Soil Conservation Service.[19]

Access to computerized data bases may be in the form of purchased searches, for instance, selective dissemination of information (SDI) services from the data base manager. Tapes of data base information may also be leased or purchased. A third approach is the purchase of time-shared access to several data bases through one vendor. In the latter case, numerous data bases can usually be searched for each request for information. Services such as this are currently available through federal agencies and commercial enterprises as well as university and other public libraries.

There is a growing use of interactive on-line retrieval systems with on-line terminal searches. In this way a user can often determine the number of responses he will get from a particular search and get some idea of the utility of his request. The following discussion outlines some of the more important data bases available to people interested in range science information.

AGRICOLA

The National Agricultural Library (now SEA-TIS) developed a family of data bases called AGRICOLA (AGRICultural On-Line Access System). This data base includes the CAtaloguing and INdexing system (CAIN), the Food and Nutrition Information and Education Materials Center information (FNIC), and the American Agricultural Economics Bibliographic data base (AGECON). AGRICOLA currently includes indexed material from over 5,000 journals with the total number of books and journal articles indexed approaching one million entries. Subjects covered by these sources of information include general agriculture, agricultural economics, agricultural products, animal industry, botany, chemistry, environmental sciences, ecology, engineering, energy in agriculture, entomology, fertilizers, foods, forestry, human nutrition, hydroponics, pesticides, plant sciences, soils, rural sociology, and water management. Most of these subjects are important to range science.

AGRICOLA can be accessed using title words of monographs and journal articles. Searches may be made by author, date of publication, type of document, geographical codes, abbreviations of journal titles, subject headings, etc. The search output will be a full biographical documentation of the literature cited, including journal title, volume number, pagination, date of publication, author(s), language of publication, and title of publication along with the National Agricultural Library (SEA-TIS) call number.

AGRICOLA can be accessed through the Technical Information Systems (TIS), Science and Education Administration (formerly National Agricultural Library) by USDA personnel. The Library Services Division of TIS will lend books to USDA personnel in response to job related requests. The general public may contact the library in person or order paid photocopies of articles from the Photocopy Section, Library Services Division, TIS (formerly Lending Division, National Agricultural Library), Beltsville, Md. 20705.[20]

At the present time on-line access to Agricola may be obtained through the following three private or commercial vendors:

1. Lockheed Information Systems, 3251 Hanover Street, Palo Alto, Calif. 94304.

2. System Development Corporation, SDS Search Service, 2500 Colorado Avenue, Santa Monica, Calif. 90406.

3. Bibliographical Retrieval Services, 1462 Erie Blvd., Schenectady, N.Y. 12305.

With the rapid growth and use of data bases, more access sources are likely to be available in the near future. Selective dissemination of information (SDI) services may also be obtained through the libraries of academic institutions.

The following information on the three data bases within AGRICOLA might be helpful in understanding its contents. The CAIN system developed by NAL is a computer-based bibliographic data system related to agricultural and allied sciences. CAIN is a collection of several data bases combined in 1970 that included the Bibliography of Agriculture and the Pesticides Documentation Bulletin. The magnetic tape of this data base is used to print many bibliographic catalogs for the NAL. It is also used for printing "Agricultural Economics," which is sold to subscribers, and by Oryx Press in producing the Bibliography of Agriculture.

FNIC is a collection of materials used by training staffs for personnel in both school and nonschool food service programs. AGECON, a collection of bibliographies containing abstracts, when available, of information on professional work currently being done by agricultural economists, is also entered into the AGRICOLA system.

CRIS

The Current Research Information System (CRIS) is a computer-based information storage and retrieval system containing information on current research activity of the U.S. Department of Agriculture (USDA), the state agricultural experiment stations (SAES), and other cooperating institutions. The design of CRIS is to provide communication among agricultural research scientists regarding current research and to provide resource managers with current coordinated information on total research programs within USDA and SAES.[21]

Information contained within CRIS includes investigators, titles of projects being conducted, performing organization(s), objectives, brief mention of methods and procedures, progress to date, and research publications produced. This information can be accessed by any scientist working within the cooperating agencies.

CRIS (originally located in USDA's Cooperative State Research Service, but now in the Agricultural Information Division, SEA-TIS) includes coverage of active and recently completed projects (those terminated within the last two years) from state agricultural experiment stations, 30 forestry schools, and other cooperating institutions. A typical information retrieval request will involve a search of some 24,000 research résumés in the CRIS data base. CRIS searches are generally broad in order to reduce the risk of missing projects of peripheral interest. Further information may be obtained by writing to Agricultural Information Division, Technical Information Systems, U.S. Dept. of Agric., Beltsville, Md. 20705.

SSIE

The Smithsonian Science Information Exchange (SSIE) established a medical science information exchange in 1949. Since that time, they have expanded this effort to include agricultural and earth sciences, as well as chemistry, engineering, biological sciences, materials and electronics, physics and mathematics, and the behavioral sciences. Currently SSIE gathers, indexes, stores and retrieves information of basic and applied work in all areas of the life and physical sciences. In the agricultural sciences alone, some 35,000 projects are processed each year.

This system is designed to complement other scientific and technical information systems available from libraries and other information centers by providing current information on active research projects before formal publication of results. Principal aims of this information system are to avoid duplication of research effort and expenditure, aid scientists in locating possible sources of support for a specific research project, provide information on support grant or contract proposals, stimulate new ideas for research planning, provide information on new experimental techniques and technological advancements, identify patterns and trends in research efforts, and to provide information on the work of specific research organizations.

SSIE provides the following information on request: supporting agency, title of project, principal investigator(s), departmental specialty doing the work, performing organization, period for which project report is viable, and a summary of the project. The report also includes the name of the supporting organization, grant contract or agency control number, and the level of funding for the project.

More than 1,300 supporting organizations provide information to SSIE, including federal agencies, state and local governments, nonprofit associations, foundations, individual research investigators, universities, colleges, and foreign organizations. Examples of some of the agencies furnishing information to SSIE are the Departments of Agriculture, Interior, and Transportation and the National Science Foundation, Environmental Protection Agency, National Oceanic and Atmospheric Administration, the Agency for International Development, Rockefeller Foundation, and state governments.

SSIE search services come in several formats. Customs searches can respond to individual requests. Such requests are formulated by individuals desiring information and transmitted to SSIE staff scientists who search the active files for Notices of Research Projects on specific subjects, specific organizations or departments performing research, specific geographic areas, or any combination of these or similar requirements. Custom searches can also include searches in the National Technical Information Service (NTIS) of the U.S. Department of Commerce. This information data base is not directly related to the agricultural and biological sciences important to range management, but it may contain some peripheral business-type information of interest to resource economics.

Research information packages are also available from SSIE. These packages of information are searches conducted by staff scientists of SSIE on subjects of current interest. Announcements of research information packages are carried in the periodical SSIE Science Newsletter. These packages are updated every three months and are offered at costs to individuals. Prices vary, depending upon the content and amount of information contained. NTIS search packages may also be combined with the SSIE information packages. The SSIE Science Newsletter is available for an annual subscription rate by contacting the address listed below.

The following are examples of research information packages that may be of interest to range scientists:

1. Strip Mine Areas and Spoil Banks: Reclamation and Vegetation: Biological, Economic and Sociological Effects.
2. Revegetation Methods and Materials for Erosion Control.
3. Wind Erosion.
4. Desert Arid, Semi-arid and Dryland Research: Agronomic and Horticultural Crop Production Under Dryland Conditions in Arid and Semi-arid Areas.
5. Remote Sensing From Aircraft or Spacecraft for Natural Resources. Environmental or Agricultural Surveys Using Aerial Photography (color, black and white, infrared), Laser Profiles, Television Observing, Spectural Signatures, Microwave Systems, and Multiband Spectral Reconaissance for Water Resources, Soil Surveys, Pollution, Meteorological Surveys, Forest or Crop Production, Plant Disease or Insect Surveys, Livestock or Wildlife Estimation.
6. World Agricultural Trade: Impact of Changes in United States Agricultural Exports.
7. Carrying Capacity of Ranges and Pastures.
8. Artificial Rumen Analysis of Feeds.
9. Appetite in Cattle, Sheep and Swine.

Selective dissemination of information (SDI) is also offered by the Exchange. A user establishes an "interest profile" and then SSIE scientists do a periodic search of the active file and forward new information to a user. Subscribers may obtain these standard SDI services and receive monthly search updates compiled automatically by computer for a single annual fee. Custom SDI services are provided on a quarterly basis.

Searches may be ordered from SSIE by telephone, cable or returning a preprinted order form. Further information may be obtained from the following address: Smithsonian Science Information Exchange Incorporated, Room 300, 1730 M Street, N.W., Washington, D.C. 20036.

SCORPIO

The Library of Congress in 1960 began to use computers to produce catalog cards, author and subject catalogs, and various indexes. The Computer Applications Office of the Library of Congress in 1973 developed an on-line information retrieval system to facilitate access to certain information by users through a relatively large number of terminals. The system developed is the Subject-Content-Oriented Retriever for Processing Information On-line (SCORPIO).[22]

SCORPIO accesses files of information, primarily bibliographic and legislative in nature. One is the Legislative Information Files that includes the contents of Digest of General Public Bills and Resolutions for the recent and current Congress. Bills and resolutions may be retrieved by bill number, public law number, sponsor(s), committee, index terms, and the bill's short title.

A Major Issue File contains concise briefs on key issues of public policy. Each brief contains a background and policy analysis, references to major current legislation, hearings, committee reports, and a chronology of the congressional action. As new issues emerge, they are added to the file. Each issue may be retrieved through SCORPIO by its number, title, and index terms.

A Citation File describes significant current periodicals, articles, Government Printing Office and UN Documents, selected Congressional Reports, lobby group publications, etc. The material is selected by the bibliographic staff of the CRS library services division based on their actual and potential research support value. This file currently contains over 100,000 citations and is increased by about 25,000 citations per year. This material can be accessed by a number code, author(s), and index terms supplied by the bibliographic staff.

The National Referral Center's Resources File contains descriptions of 10,000 organizations qualified and willing to answer questions on virtually any topic in science and technology. This file is a collection of records from the Information Resources Information System (IRIS) and from "A Directory of Information Resources in the United States." Again, these resources may be tapped by accession number, organization, index terms, and geographic descriptors.

Another file in the SCORPIO collection is the Selected Science and Technology File which contains records on about 90,000 references selected from the Library of Congress's Machine Readable Catalogue data base (MARC). This file describes English books and catalogues published since 1969, and it is keyed to the Science Reading Room. About 6,000 titles including journals, handbooks, and scientific encyclopedias in many languages are included. The citation may be retrieved by Library of Congress card number, Library of Congress classification number, title, author, and index terms.

Users can be trained in three aspects of SCORPIO that include terminal operation, file content, and SCORPIO commands that allow users to access the system. Consequently, access to terminals allows users to become proficient and independently able to access SCORPIO. A training session requires about 45 minutes, after which users can conduct their own searches and aid other people in using this system.

For further information on the SCORPIO system contact: Library of Congress, Administrative Department, Information Systems Office, Computer Applications Office, Washington, D.C. 20540.

DIALOG

One of the largest and most comprehensive collections of on-line data bases operated by commercial vendors is the Dialog Service of the Lockheed Information Systems.[23] The more than 90 separate data bases within this system cover a wide range of subject matter in science, engineering, social sciences, business, and economics. These data bases, which include most of the ones mentioned in previous sections, are regularly updated to include the most recent information.

Dialog makes available special search features to the research community. These features are:

1. Selective dissemination of information (SDI) services which allow a user access to several data bases at one time to meet specific requests.
2. Full-text searching may be done in which the computer-stored records, including the subject identifiers (index terms), titles, and abstract, are searched for a particular word or phrase. The following is a partial list of the data bases and the suppliers of data bases to the Dialog Service:

DATA BASE	SUPPLIER
1. ERIC	Educational Resources Information Center
2. CA CONDENSATES 1970-1971	American Chemical Society
3. CA CONDENSATES 1972-1976	American Chemical Society
4. CA CONDENSATES/CASIA 1977-Present	American Chemical Society
5. BIOSIS PREVIEWS	Biosciences Information Service of Biological Abstracts, Incorporated
6. NTIS	National Technical Information Service, U.S. Dept. of Commerce
7. AGRICOLA	SEA-Technical Information Systems (formerly National Agricultural Library), U.S. Dept. of Agriculture
8. FOUNDATION DIRECTORY	The Foundation Center

9. FOUNDATION GRANTS	The Foundation Center
10. SCISEARCH	Institute for Scientific Information
11. COMPREHENSIVE DISSERTATION ABS	Xerox University Microfilms
12. COMMONWEALTH AGRICULTURAL BUREAU ABSTRACTS	Commonwealth Agricultural Bureaux
13. FOOD SCIENCE AND TECHNOLOGY ABSTRACTS	International Food Information Service
14. CURRENT RESEARCH INFORMATION SYSTEM	USDA
15. SSIE CURRENT RESEARCH	Smithsonian Science Information Exchange

Copies of the Dialog subject indexes can be obtained for aiding users in developing search strategies and in comparing data base content. To access efficiently any information system, the user needs to understand the content of the data bases. Access to the system may be through compatible terminals owned or leased by the user of a dial-up terminal that uses either a TYMNET or TELENET data communications network services. Users may direct dial the Dialog Computer Center in Palo Alto, Calif., for easy telephone access.

A wide range of compatible terminals is available to the user. These can be leased for a typical cost of $85 to $150 per month depending upon the brand selected. Terminals of many brands of computers are adaptable to the Lockheed Information System.

More information on the Dialog system may be obtained by writing to: Lockheed Information, Code 5020/201, 3251 Hanover Street, Palo Alto, Calif. 94304.

Dialog representatives are also located in the following cities:

Lockheed Information Systems
Suite 201
Crawford Savings Bldg.
1400 Summit Avenue
Oak Brook, Terrace, Ill. 60181

Lockheed Information Systems
200 Park Avenue
Suite 303 East
New York, N.Y. 10017

Lockheed
900 17th St., N.W.
Washington, D.C. 20006

Learned Information (Europe) Ltd.
Woodside House
Hinksey Hill
Oxford OX1 5BP, UK.

ORBIT

The System Development Corporation's ORBIT provides on-line or interactive searching of several data bases of interest to range managers. At present these data bases are duplicates of those available through the Lockheed Information Systems. The data bases accessable through the SDC search services include science and technology, business, and social sciences. The data bases applicable to the field of range management are CAIN, Chemical Abstracts, Condensates, NTIS and SSIE.[24]

The system is accessed through modern telecommunications and the use of a terminal telephone and a user identification number provided by SDC. Searches are made by selecting a term or a series of terms for searching a particular subject matter area. If a printer terminal is available, part or all of the citations can be printed directly to the user's terminal. Otherwise, the system can be keyed to print the results of a search, and these will be airmailed to the user on the day following the request.

SDC has been involved in computer-based information systems since 1956. Since 1974, SDC search services have been available internationally as well as in this country. This system can be accessed by inexpensive long-distance communications through the nationwide TYMSHARE Communications NetWork (TYMNET) or through TELENET. For further information contact SDC Search Service at:

2500 Colorado Avenue
Santa Monica, Calif. 90406

7929 Westpark Drive
McLean, Va. 22101

P.O. Box 22
Paramus, N.J. 07652

CURRENT AWARENESS

A supplementary approach to searching range science and related literature is current awareness, i.e. ongoing efforts to keep updated on what is new in the field, who is doing it, where it is being done, and its application. This approach is particularly helpful to range scientists and others whose literature needs are continual and concentrated in the range science area. To be effective, current awareness must be based on a regular, priority program of reviewing the current literature being published in the field.

Suggestions for using the published literature of range science for accelerating current awareness include the following:

1. Read the journals, abstracts or proceedings of papers at annual meetings, and special publications of pertinent professional societies.
2. Systematically review articles in a selected core of additional primary periodicals and monographic series.

3. Systematically scan secondary periodicals and monographic series and select articles of special interest for review; this can be expedited by use of "Current Contents."

4. Examine news and notes, new publications, and current literature sections in selected journals; most journals include book reviews and notes on new publications.

5. Read on a systematic basis selected general, review-type publications on range science; these may include new textbooks, "status-of-our-knowledge" articles or monographs, and symposia proceedings.

6. Use prepared bibliographies and even abstracting and indexing journals in searching for new publications.

7. Place your name on mailings to receive new lists of available publications as issued.

8. Use current awareness computer searches.

9. Maintain a personal library to support and provide greater depth to current awareness efforts.

CURRENT CONTENTS: AGRICULTURE, BIOLOGY, AND ENVIRONMENTAL SCIENCES is one of several similar weekly information periodicals published by the Institute for Scientific Information, 325 Chestnut St., Philadelphia, Pa. 19106. Begun in 1970 and presently (1979) in Volume 10, it covers over 1,030 U.S. and foreign journals and selected serials, including food and veterinary sciences. This periodical reproduces the table of contents pages and makes them available about the time of publication of the journals or serials; it also provides the names and addresses of the authors for use in requesting reprints. Current Contents is suggested for literature scanning, reading selection, and rapid dissemination of information; it also now has a "Current Book Contents" section.

Current awareness computer search services have recently become available. Frequent, regular searches are made on a fee basis based on a customized profile of subjects in the user's field of interest. Reports comprising selective bibliographies or even abstracts or information summaries are regularly provided for subscribers. In addition to selective dissemination of information (SDI) services from DIALOG or ORBIT, two other such services are offered by Bioscience Information Services of Biological Abstracts and by the Institute of Scientific Information. The considerable cost and incomplete coverage of range subjects have minimized the use of such services by range scientists; at best, current awareness computer search services should be considered as an assist and not a replacement of manual approaches to current awareness.

New bibliographies, particularly when including informative abstracts, may be used for current awareness. No annual review or advancements series has been or is presently being published for the field of range science as in related fields although such has been proposed. Such an annual review of progress specific to range science and management is much needed to supplement the infrequent review articles in the Journal of Range Management and occasional "sciential" papers published by the Society for Range Management.

Current literature sections are very useful for current awareness but are presently included only in a few periodicals, i.e. Journal of Forestry, Journal of Soil and Water Conservation, and Wildlife Society Bulletin. Such sections were previously included in the Journal of Range Management (1949-1968) and Weed Science (Vol. 1-10) but were later discontinued because, in part, of problems in rapid manual preparation and space considerations in these journals. However, the advantages of current literature sections to practitioners and field personnel in range management probably have been underrated.

Chapter 3
THE LITERATURE OF RANGE SCIENCE

LITERATURE USED BY RANGE SCIENTISTS

The literature of range science is voluminous but widely scattered through an assortment of journal articles, monographs, monographic series, symposium proceedings, etc. In a recent bibliography of North American range management literature, about 20,600 literature items published between 1935 and 1977 were selected as being representative of range science and providing substantial subject matter information.[25] A similar bibliography covering the period prior to 1935 included 8,274 literature entries.[26] However, both publications were designed to be more selective than exhaustive.

Literature citations made by authors in the Journal of Range Management in 1976 and 1977 have been used to indicate and evaluate the various sources of range science literature.[27] The relative use made of the various forms of literature was as follows (1976-77):

	Percent
Monographs	14.6
Books	13.0
Pamphlets	1.7
Serials	77.4
Periodicals	51.2
Monographic series (Federal)	10.6
Monographic series (State)	8.3
Regular proceedings, annuals, reviews	7.3
Unpublished literature	6.4
Dissertations and theses	3.8
Unpublished manuscripts	2.6
Personal communications	0.1
Miscellaneous	1.6

This shows that about half of current literature citations in the Journal of Range Management (1976-77) was taken from periodical literature, with total serial literature comprising about three-fourths. Monographic series (18.9 percent total), books (13.0 percent), and regular proceedings, annuals, and reviews

(7.3 percent) ranked second, third, and fourth. Dissertations and theses (3.8 percent), unpublished manuscripts (2.6 percent), and pamphlets (1.7 percent) were used rather infrequently, and personal communications (0.1 percent) were essentially not used as documented sources of information. It is probable that high use of periodical literature has resulted both from its popularity and its more effective indexing and abstracting. The increasing use of annual transactions, regular and irregular symposia proceedings, and books also was noted.

The relative frequency with which authors in the Journal of Range Management cited the various periodicals during 1976-77 was as follows:

	Percent of all periodical citations (1976-1977)
Journal of Range Management	27.8
Ecology	6.9
Weed Science	6.9
Agronomy Journal	5.0
Journal of Wildlife Management	4.8
Journal of Animal Science	3.5
Journal of Forestry	2.2
Journal of the British Grassland Society	1.9
Ecological Monographs	1.8
Soil Science Society of America Journal	1.7
Soil Science	1.4
American Journal of Veterinary Research	1.1
Crop Science	1.1
Canadian Journal of Plant Science	1.0

The six most cited journals provided 54.9 percent of the periodical citations made in the Journal of Range Management, 1976-77. Journals comprising 0.9 to 0.7 percent each were Phytochemistry, Journal of Ecology, Plant Physiology, Australian Journal of Experimental Agriculture and Animal Husbandry, Canadian Journal of Animal Science, Botanical Gazette, American Journal of Botany, Northwest Science, Botanical Review, American Midland Naturalist, and Journal of Agricultural Research.

MAJOR PERIODICALS AND SERIALS

Based on their contents and frequency of citation by range science authors, the following are considered as principal periodicals and serials for range science information:

AGRONOMY JOURNAL (formerly Journal of the American Society of Agronomy)
Publisher: American Society of Agronomy, 677 S. Segoe Road, Madison, Wisc. 53711.
Publication: began 1907; present volume (1979), Vol. 71; issued bimonthly (1959 to present), monthly (1921-58), nine times per year (1915-20), monthly (1911-14), and annually (1907-10).

Evaluation: reports research on agronomy, crop science, and soil science; considerable emphasis given to pasture and harvested roughages; infrequent articles on rangelands, but more prior to about 1960; includes book reviews; cumulative index for Vols. 1-50.

AMERICAN SOCIETY OF ANIMAL SCIENCE, WESTERN SECTION PROCEEDINGS
Publisher: host universities for ASAS Western Section (contact American Society of Animal Science, 113 North Neil St., Champaign, Ill. 61820).
Publication: began 1950; present volume (1979), Vol. 30; issued annually.
Evaluation: papers on all phases of animal production, including range and pasture forages and range livestock production.

AUSTRALIAN RANGELAND JOURNAL
Publisher: Australian Rangeland Society, c/o CSIRO, Wembley, Western Australia.
Publication: began 1977 with Vol. 1; two issues per year projected.
Evaluation: emphasis given to methods of rangeland management, range education and extension, new concepts or objectives for research, research results, integration of economic imperatives, and multiple use of range resources; book reviews included.

CANADIAN JOURNAL OF ANIMAL SCIENCE
Publisher: Agricultural Institute of Canada, 151 Slater Street, Ottawa, Ont. K1P 5H4 (for the Canadian Society of Animal Science).
Publication: began 1957 with Vol. 37 (formerly part of Canadian Journal of Agricultural Science); present volume (1979), Vol. 59; issued semiannually (1957-63), every four months (1964-71), quarterly (1972 to present).
Evaluation: includes articles on animal breeding and genetics; ruminant and monogastric nutrition; animal physiology, diseases, care, and management; articles frequently included on range animal nutrition and production.

CANADIAN JOURNAL OF PLANT SCIENCE
Publisher: Agricultural Institute of Canada, 151 Slater Street, Ottawa, Ont. K1P 5H4 (for the Canadian Society of Agronomy and the Canadian Society for Horticultural Science).
Publication: began 1957 with Vol. 37 (formerly part of Canadian Journal of Agricultural Science); present volume (1979), Vol. 59; issued quarterly (1957-63), bimonthly (1964-72), and quarterly (1973 to present).
Evaluation: emphasizes articles on forages, grain production, fruit and vegetables, and weeds; includes many articles on range improvement and management; includes announcements of registration of plant cultivars in Canada.

CROP SCIENCE
Publisher: Crop Science Society of America, 677 S. Segoe Road, Madison, Wisc. 53711.
Publication: began 1961; present volume (1979), Vol. 19; issued bimonthly.
Evaluation: reports of research in field crops, including genetics, breeding, physiology, and ecology; includes registration of crop cultivars for United States; articles on pasture and harvested forage plants included.

ECOLOGICAL MONOGRAPHS
Publisher: Duke University Press, Durham, N.C. 27708 (Ecological Society of America, owner).
Publication: began 1931; present volume (1979), Vol. 49; issued quarterly.
Evaluation: comprehensive articles on all aspects of ecology including range ecology; 20-year indexes for Vols. 1-20 (1931-50) and Vols. 21-40 (1951-70).

ECOLOGY
Publisher: Duke University Press, Durham, N.C. 22708 (Ecological Society of America, owner).
Publication: began 1920; present volume (1979), Vol. 60; issued quarterly (1920-66), bimonthly (1967 to present).
Evaluation: articles on all phases of ecology, i.e., all forms of life in relation to their environment; much included on range ecology; book reviews included; cumulative indexes, Vols. 1-30 and 31-50.

GRASS AND FORAGE SCIENCE (formerly Journal of the British Grassland Society, pub. 1946-79)
Publisher: British Grassland Society, Hurley, Maidenhead, Berks SL6 5LR, UK.
Publication: began 1979; present volume (1979), Vol. 34; issued quarterly.
Evaluation: emphasis given to the production, management, and utilization of pasture and harvested forages in temperate regions; rangeland given minimal emphasis; book reviews.

INTERNATIONAL GRASSLAND CONGRESS PROCEEDINGS
Publisher: hosting national government (contact Liaison Officer, Pasture and Forage Crops Group, FAO, Rome, Italy, or Grassland Research Institute, Hurley, Maidenhead, Berks SL6 5LR, UK.)
Year began: 1927; congress now held; proceedings issued about every two years.
Evaluation: papers dealing with the production, improvement, management, and use of grasslands, including harvested forage crops, temporary and permanent pastures, and rangelands.

INTERNATIONAL RANGELAND CONGRESS PROCEEDINGS
Publisher: Society for Range Management, 2760 West Fifth Ave., Denver, Colo. 80204.
Year began: first congress, August 14-18, 1978, at Denver, Colo.; projected periodically.
Evaluation: papers dealing with all phases of range science throughout the world.

JOURNAL OF ANIMAL SCIENCE
Publisher: American Society of Animal Science, 113 North Neil St., Champaign, Ill. 61820.
Publication: began 1942; present volumes (1979), Vols. 48 and 49 (two volumes per year beginning in 1970); issued quarterly (1942-66), bimonthly (1967-68), and monthly (1969 to present).
Evaluation: emphasizes applied animal science, breeding and genetics, meat science and muscle biology, ruminant and nonruminant nutrition, and physiology and endocrinology; includes numerous articles on range animal nutrition and management; news and notes; cumulative indexes, Vol. 1-24 and 25-34.

JOURNAL OF FORESTRY (formerly Forestry Quarterly prior to 1915)
Publisher: Society of American Foresters, 5400 Grosvenor Lane, Washington, D.C. 20014.
Publication: began 1902; present volume (1979), Vol. 77; issued quarterly (Vol. 1-14), eight per year (Vol. 15-31), monthly (Vol. 32 to present).
Evaluation: devoted to advancing the science, technology, practice and education of professional forestry; includes articles on forest range, plant ecology, and wildlife (many articles on range management prior to 1950); special features include current literature (including range), publications of interest, and book reviews.

JOURNAL OF RANGE MANAGEMENT
Publisher: Society for Range Management, 2760 West Fifth Ave., Denver, Colo. 80204.
Publication: began 1948; present volume (1979), Vol. 32; issued bimonthly (1951-present), quarterly (1949-50), annually (1948).
Evaluation: covers all phases of the study, management, and use of rangeland ecosystems and resources; the principal periodical for range science literature; special features include book reviews, new publications, current literature (1949 to 1968 only); 10-year indexes, Vols. 1-10 and 11-20; 30-year cumulative index projected for publication in 1979.

JOURNAL OF SOIL AND WATER CONSERVATION
Publisher: Soil Conservation Society of America, 7515 Northeast Ankeny Road, Ankeny, Iowa 50021.
Publication: began 1946; present volume (1979), Vol. 34; issued bimonthly.
Evaluation: emphasizes the principles of good land use; includes all aspects of management and conservation of renewable natural resources; book reviews and current literature; cumulative index, Vols. 1-20.

JOURNAL OF WILDLIFE MANAGEMENT
Publisher: The Wildlife Society, Inc., 7101 Wisconsin Ave., N.W., Washington, D.C. 20014.
Publication: began 1937; present volume (1979), Vol. 43; quarterly.
Evaluation: covers the field of wildlife biology and wildlife management, including many articles related to rangelands and big game; book reviews and recent books; cumulative indexes, Vols. 1-10, 11-20, 21-30.

RANGELANDS (replaced Rangeman's News, pub. 1969-74; name changed from Rangeman's Journal, pub. 1974-78).
Publisher: Society for Range Management, 2760 West Fifth Ave., Denver, Colo. 80204.
Publication: began 1974 as Rangeman's Journal; present volume (1979), Vol. 1; issued bimonthly (1975 to present), semiannually (1974).
Evaluation: the nontechnical counterpart of the Journal of Range Management; includes articles, news and notes, governmental affairs, and new publications announcements; emphasis given to "facts, ideas, and philosophies pertaining to rangelands and their resources, uses, study, management, and practices."

SOCIETY FOR RANGE MANAGEMENT, ABSTRACTS OF PAPERS OF THE ANNUAL MEETINGS
Publisher: Society for Range Management, 2760 W. Fifth Ave., Denver, Colo. 80204.
Publication: issued annually (except 1975); published separately.
Evaluation: abstracts of papers presented at the Society's annual meetings covering all phases of range science.

SOIL SCIENCE SOCIETY OF AMERICA JOURNAL (formerly Proceedings)
Publisher: Soil Science Society of America, 677 S. Segoe Rd., Madison, Wis. 53711.
Publication: began in 1936 with Vol. 1 (published in 1937); present volume (1979), Vol. 43; bimonthly.
Evaluation: covers all aspects of soil science, fertility, soil and water management and conservation, including forest and range soils; book reviews and soil briefs.

WEED SCIENCE (formerly Weeds)
Publisher: Weed Science Society of America, 113 N. Neil St., Champaign, Ill. 61820.
Publication: began 1952; present volume (1979), Vol. 27; issued bimonthly (Vol. 18 to present), quarterly (Vols. 1-17).
Evaluation: emphasis given to ecology and control of weeds including undesirable range plants; book reviews; bibliography of recent weed literature included through Vol. 10.

RELATED PERIODICALS AND SERIALS

The following technical and semitechnical periodicals and serials include occasional articles on North American rangelands or articles pertaining to or applicable in the broad field of range science and management:

ADVANCES IN AGRONOMY
Publisher: Academic Press, 111 Fifth Ave., New York, N.Y. 10003 (prepared under the auspices of the American Society of Agronomy)
Publication: began 1949; present volume (1979), Vol. 31; issued annually; cumulative author and subject indexes for Vols. 1-15 (1949-1963) in Vol. 16.
Evaluation: review papers on all phases of crops and soils, including the production and utilization of pastures and harvested roughages.

ADVANCES IN ECOLOGICAL RESEARCH
Publisher: Academic Press, 111 Fifth Ave., New York, N.Y. 10003.
Publication: began 1962; Vol. 9 (1975); issued irregularly.
Evaluation: reviews on ecology, mostly on basic science topics with worldwide implication.

AGRICULTURAL RESEARCH
Publisher: Science and Education Administration (formerly by Agricultural Re-

search Service), USDA, Washington, D.C. 20250.
Publication: began 1953; present volume (1979), Vol. 28 (July 1978-June 1979); issued monthly.
Evaluation: short, prepublication articles on agricultural research including range research by SEA-AR; also includes Agrisearch Notes; cumulative indexes issued about every three years.

AMERICAN JOURNAL OF AGRICULTURAL ECONOMICS (formerly Journal of Farm Economics)
Publisher: American Agricultural Economics Association, University of Kentucky, Lexington, Ky. 40506.
Publication: began 1919; present volume (1979), Vol. 61; now issued five times per year.
Evaluation: covers broad field of agricultural economics including economic theory, production economics, farm-ranch management, agricultural marketing, and natural resource economics; book reviews; lists new books received.

AMERICAN JOURNAL OF BOTANY
Publisher: Botanical Society of America, c/o New York Botanical Garden, Bronx, N.Y. 10458.
Publication: began 1914; present volume (1979), Vol. 66; now issued 10 times per year.
Evaluation: covers all branches of the basic plant sciences.

AMERICAN JOURNAL OF VETERINARY RESEARCH
Publisher: American Veterinary Medical Association, 930 N. Meacham Road, Schaumburg, Ill. 60196.
Publication: began 1940; present volume (1979), Vol. 40; issued monthly.
Evaluation: reports research on diseases and nutrition of domestic, wild, and fur-bearing animals; includes animal poisoning by plants.

AMERICAN MIDLAND NATURALIST
Publisher: University of Notre Dame, Notre Dame, Ind. 46556.
Publication: began 1909; present volumes (1979), Vols. 101-102; issued quarterly (two issues per volume).
Evaluation: emphasis given to broad field of biology but includes articles on range ecology; cumulative index, Vols. 1-60.

AMERICAN SOCIETY OF FARM MANAGERS AND RURAL APPRAISERS JOURNAL
Publisher: American Society of Farm Managers and Rural Appraisers, P.O. Box 6857, Denver, Colo. 80206.
Publication: began 1937; present volume (1979), Vol. 43; presently issued semi-annually.
Evaluation: emphasis given to management and appraisal of rural properties, modern farm-ranch management practices, and agricultural finance; includes book reviews.

ANTELOPE STATES WORKSHOPS PROCEEDINGS
Publisher: host state fish and game departments (First, 1965, Santa Fe; Second,

1966, Denver; Third, 1968, Casper; Fourth, 1970, Scottsbluff; Fifth, 1972, Billings; Sixth, 1974, Salt Lake City; Seventh, 1976, Twin Falls).
Publication: began 1965; mostly issued bienially.
Evaluation: papers on antelope biology, ecology, management, and research in western North America; includes business of the workshop also.

AUSTRALIAN JOURNAL OF AGRICULTURAL RESEARCH
Publisher: Commonwealth Scientific and Industrial Research Organization, Melbourne, Victoria, Australia.
Publication: began 1950; present volume (1979), Vol. 30; issued bimonthly (Vols. 6-present), quarterly (Vols. 1-5).
Evaluation: physical, chemical, and/or biological aspects of agricultural systems, including rangelands pertinent to Australian conditions.

AUSTRALIAN JOURNAL OF EXPERIMENTAL AGRICULTURE AND ANIMAL HUSBANDRY
Publisher: Commonwealth Scientific and Industrial Research Organization, Melbourne, Victoria, Australia (for the Australian Agriculture Council).
Publication: began 1961; present volume (1979), Vol. 19; issued bimonthly (Vols. 7-present), quarterly (Vols. 1-6).
Evaluation: covers the broad field of Australian agriculture including management of range lands and livestock.

BOTANICAL GAZETTE
Publisher: University of Chicago Press, Chicago, Ill. 60637.
Publication: began 1875; present volume (1979), Vol. 140; issued quarterly.
Evaluation: emphasizes botany and plant sciences generally, including morphology, physiology, and ecology.

BOTANICAL REVIEW
Publisher: New York Botanical Garden, Bronx, N.Y. 10458.
Publication: began 1935; present volume (1979), Vol. 45; issued quarterly.
Evaluation: review articles interpreting progress in all areas of botanical sciences.

CALIFORNIA AGRICULTURE
Publisher: University of California, Division of Agricultural Science, Berkeley, Calif. 94720.
Publication: began 1947; present volume (1979), Vol. 33; issued monthly.
Evaluation: emphasizes research progress in agriculture, including range management, by the California Agricultural Experiment Station and Cooperative Extension Service.

CALIFORNIA FISH AND GAME
Publisher: California Department of Fish and Game, 1416 Ninth St., Sacramento, Calif. 95814.
Publication: began 1914; present volume (1979), Vol. 33; issued monthly.
Evaluation: includes technical and popular articles on the biology, ecology, conservation, and management of wildlife in California; book reviews.

CANADIAN FARM ECONOMICS
Publisher: Economics Branch, Canada Department of Agriculture, Ottawa, Ont.
Publication: began 1966; present volume (1979), Vol. 14; issued bimonthly.
Evaluation: emphasis given to current information on Canadian agricultural economics for economists, extension, education, agribusiness, leading farmers, etc.; recent publications section included.

CANADIAN JOURNAL OF AGRICULTURAL ECONOMICS
Publisher: Canadian Agricultural Economics Society, Ottawa, Ont.
Publication: began 1952; present volume (1979), Vol. 27; issued five times per year (three numbered issues, proceedings of the annual workshop, and proceedings of the annual meeting).
Evaluation: emphasis given to the broad field of Canadian agricultural economics, including farm-ranch management.

CANADIAN JOURNAL OF AGRICULTURAL SCIENCE (formerly Scientific Agriculture, 1921-1952)
Publisher: Agricultural Institute of Canada, 151 Slater St., Ottawa, Ont. K1P 5H4.
Publication: began 1921, issued bimonthly; discontinued in 1956 with Vol. 36; continued as Canadian Journal of Animal Science, Canadian Journal of Plant Science, and Canadian Journal of Soil Science.
Evaluation: reported scientific and technical articles on Canadian agriculture, including articles on range and pasture management.

CANADIAN JOURNAL OF SOIL SCIENCE
Publisher: Agricultural Institute of Canada, 151 Slater St., Ottawa, Ont. K1P 5H4 (for the Canadian Society of Soil Science).
Publication: began in 1957 as Vol. 37 (formerly included in Canadian Journal of Agricultural Science); present volume (1979), Vol. 59; issued quarterly (Vol. 53-present), semiannually (Vols. 37-43), and three times per year (Vols. 44-52).
Evaluation: reports original research on soils, including soil analysis and chemistry, amendments, management, and environmental influences, and their relationship to food production.

CROPS AND SOILS
Publisher: American Society of Agronomy, 677 S. Segoe Road, Madison, Wisc. 53711.
Publication: began 1948; present volume, Vol. 31 (October 1978-August 1979); issued nine times per year.
Evaluation: applied research reports on crop production and soil management for commercial farmers, educators, and agribusiness; includes research briefs, new publications, and book reviews.

DESERT BIGHORN COUNCIL TRANSACTIONS
Publisher: Desert Bighorn Council, Death Valley National Monument, Death Valley, Calif. 92328.
Publication: began prior 1957; issued annually.

Evaluation: papers on desert bighorn biology, ecology, management, and research in western North America; also includes reports and minutes of planning meetings.

DOWN TO EARTH
Publisher: Dow Chemical Co., Midland, Mich. 48640.
Publication: began 1948; present volume, Vol. 34 (summer 1978-spring 1979); issued quarterly.
Evaluation: status and progress in use of agricultural chemicals, including herbicides and insecticides.

FOREST SCIENCE
Publisher: Society of American Foresters, 5400 Grosvenor Lane, Washington, D.C. 20014.
Publication: began 1955; present volume (1979), Vol. 25; issued quarterly.
Evaluation: emphasizes research and technological progress in forest science; some articles pertain to forest soils and ecology, watershed conservation, and forest range.

GREAT BASIN NATURALIST
Publisher: Brigham Young University Press, Provo, Utah 84602.
Publication: began 1939; present volume (1979), Vol. 39; issued quarterly (occasionally 1-3 issues per year).
Evaluation: articles emphasize the biology and natural history of North America; includes infrequent articles on range ecology; cumulative index, Vols. 1-30.

HILGARDIA
Publisher: California Agricultural Experiment Station, Berkeley, Calif. 94720.
Publication: began 1924; present volume (1979), Vol. 47; various numbers issued per volume.
Evaluation: technical articles on California agriculture.

JOURNAL OF AGRICULTURAL RESEARCH
Publisher: U.S. Government Printing Office, Washington, D.C., for the USDA and the Association of Land Grant Colleges and Universities.
Publication: began 1913; issued irregularly (mostly 12 issues per volume in later years); discontinued with Vol. 77 on 15 Dec. 1948).
Evaluation: published the results of agricultural research at state and federal levels; covered broad field of agriculture, including range management.

JOURNAL OF AGRONOMIC EDUCATION
Publisher: American Society of Agronomy, 677 S. Segoe Road, Madison, Wisc. 53711.
Publication: began 1972; present volume (1979), Vol. 8; issued annually; cumulative subject and author indexes for Vols. 1-5 (1972-1976) in Vol. 6; book reviews.
Evaluation: original research and review papers on resident, extension, and industry education in plant and soil sciences.

JOURNAL OF APPLIED ECOLOGY
Publisher: British Ecological Society, Reading, England.
Publication: began 1964; present volume (1979), Vol. 16; three issues per year.
Evaluation: all phases of plant ecology of general interest; occasional article on Canadian or U.S. range ecology; book reviews and short notices included.

JOURNAL OF DAIRY SCIENCE
Publisher: American Dairy Science Association, 113 N. Neil St., Champaign, Ill. 61820.
Publication: began 1917; present volume (1979), Vol. 62; issued monthly.
Evaluation: emphasizes all phases of dairy production; includes articles on production and utilization of cultivated pastures and harvested forages; book reviews.

JOURNAL OF ECOLOGY
Publisher: British Ecological Society, Reading, England.
Publication: began 1913; present volume (1979), Vol. 67; issued three times per year.
Evaluation: covers all phases of bioecology; occasional article on Canadian or U.S. range ecology; book reviews; cumulative indexes, Vols. 1-20, 21-50.

JOURNAL OF ECONOMIC ENTOMOLOGY
Publisher: Entomological Society of America, 4603 Calvert Road, College Park, Md. 20740.
Publication: began 1908; present volume (1979), Vol. 72; now issued bimonthly.
Evaluation: emphasis given to controlling crop and animal insects and their relationship to agricultural production; some articles on rangeland insects.

JOURNAL OF ENVIRONMENTAL QUALITY
Publisher: American Society of Agronomy, 677 S. Segoe Road, Madison, Wisc. 53711 (in cooperation with Crop Science Society of America and Soil Science Society of America).
Publication: began 1972; present volume (1979), Vol. 8; issued quarterly.
Evaluation: emphasizes environmental quality in natural and agricultural ecosystems; book reviews.

JOURNAL OF MAMMALOGY
Publisher: American Society of Mammalogists, Oklahoma State University Museum, Stillwater, Okla. 74074.
Publication: began 1919; present volume (1979), Vol. 60; issued quarterly.
Evaluation: covers all phases of mammalogy; selected articles pertain to the biology of rodents, upland game animals, or big game animals; book reviews; cumulative indexes published every 10 years.

JOURNAL OF THE AMERICAN VETERINARY MEDICAL ASSOCIATION
Publisher: American Veterinary Medical Association, 930 N. Meacham Road, Schaumburg, Ill. 60196.

Publication: began 1852; present volumes (1979), Vols. 174 and 175; presently issued semimonthly (12 issues per volume).
Evaluation: covers diagnosis and treatment of animal diseases, reproduction and nutrition problems, and clinical items; selected articles on livestock poisoning by plants; book reviews.

NEBRASKA FARM, RANCH AND HOME QUARTERLY (formerly Nebraska College of Agriculture Quarterly)
Publisher: Institute of Agriculture and Natural Resources, University of Nebraska, Lincoln, Neb. 68508.
Publication: began 1952; present volume (1979), Vol. 26; issued quarterly.
Evaluation: covers all phases of agriculture and related sciences in Nebraska, including semitechnical articles on range livestock and range forage production and utilization.

NEW ZEALAND JOURNAL OF AGRICULTURAL RESEARCH
Publisher: New Zealand Department of Scientific and Industrial Research, Wellington, N.Z.
Publication: began 1958; present volume (1979), Vol. 22; issued quarterly.
Evaluation: covers all aspects of agriculture with New Zealand application.

NORTH DAKOTA FARM RESEARCH (formerly North Dakota Agricultural Experiment Station Bimonthly Bulletin)
Publisher: North Dakota Agricultural Experiment Station, Fargo, North Dakota 58102.
Publication: began 1938; present volume, Vol. 36 (September 1978-August 1979); issued bimonthly.
Evaluation: progress reports of research on North Dakota agriculture; commonly includes articles on pasture and range management.

NORTHWEST SCIENCE
Publisher: Northwest Scientific Association, Pullman, Washington 99163.
Publication: began 1927; present volume (1979), Vol. 53; issued quarterly.
Evaluation: original and review papers on research in the physical, biological, and social sciences; articles on range ecology frequently included; cumulative index, Vols. 21-46.

PLANT PHYSIOLOGY
Publisher: American Society of Plant Physiologists, 9650 Rockville Pike, Bethesda, Md. 20014.
Publication: began 1926; present volumes (1979), Vols. 62-63; now issued monthly (two volumes per year).
Evaluation: primary research on plant physiology and biochemistry; occasional articles with forage plant emphasis.

PROCEEDINGS OF THE SOCIETY OF AMERICAN FORESTERS
Publishers: Society of American Foresters, 5400 Grosvenor Lane, Washington, D.C. 20014.
Publication: began 1906; present issue, 1978 annual volume; issued annually.

Evaluation: papers on the science, technology, and practice of professional forestry; includes some papers on forest range ecology and management.

PROGRESSIVE AGRICULTURE IN ARIZONA
Publisher: College of Agriculture, University of Arizona, Tucson, Ariz. 85721.
Publication: began 1949; present volume (1979), Vol. 31; issued quarterly.
Evaluation: nontechnical reports of research and services pertaining to Arizona agriculture; includes a moderate number of articles on rangelands and range livestock.

SOIL CONSERVATION
Publisher: Soil Conservation Service, USDA, Washington, D.C. 20250.
Publication: began 1935; present volume (1979), Vol. 44; issued monthly.
Evaluation: applied articles on soil conservation including range conservation; book reviews.

SOIL SCIENCE
Publisher: Williams and Watkins Co., Baltimore, Md. 21202.
Publication: began 1916; present volumes (1979), Vols. 127-28; now issued monthly (six issues per volume).
Evaluation: all phases of soil science; occasional articles pertain to cultivated pasture and rangeland; book reviews.

SOUTH DAKOTA FARM AND HOME RESEARCH
Publisher: South Dakota Agricultural Experiment Station, Brookings, South Dakota 57006.
Publication: began 1950; present volume (1979), Vol. 30; issued quarterly.
Evaluation: progress reports of research on South Dakota agriculture; commonly includes articles on cultivated pasture and range management.

SOUTHWESTERN NATURALIST
Publisher: The Southwestern Association of Naturalists, Austin, Texas.
Publication: began 1953; present volume (1979), Vol. 24; issued quarterly.
Evaluation: natural history studies of plants and animals in the southwestern states; frequently includes articles on range ecology; book reviews.

TALL TIMBERS FIRE ECOLOGY ANNUAL CONFERENCE PROCEEDINGS
Publisher: Tall Timbers Research Station, Route 1, Box 110, Tallahassee, Fla. 32302.
Publication: began 1962; Vol. 15 (1974, pub. 1976); issued annually.
Evaluation: papers on the ecology of fire and the use of burning as a land management tool.

TEXAS AGRICULTURAL PROGRESS
Publisher: Department of Agricultural Communications, Texas A&M University, College Station, Texas 77843.
Publication: began 1955; present volume (1979), Vol. 25; issued quarterly.
Evaluation: semitechnical and popular extension and research reports on Texas

agriculture; articles on cultivated pasture and range forages, lands, and animals frequently included; also lists recent Texas agricultural publications.

TRANSACTIONS OF THE NORTH AMERICAN WILDLIFE AND NATURAL RESOURCES CONFERENCE
Publisher: Wildlife Management Institute, 709 Wire Bldg., Washington, D.C. 20005.
Publication: began 1915; numbered volumes began with Vol. 1 (1936); present volume (1979), Vol. 44; issued annually; 40-year cumulative index.
Evaluation: papers on the ecology, management, and regulations relative to North American wildlife and related natural resources.

UTAH SCIENCE (formerly Utah Farm and Home Science)
Publisher: Utah Agricultural Experiment Station, Logan, Utah 84322.
Publication: began 1940; present volume (1979), Vol. 40; issued quarterly.
Evaluation: articles primarily related to research in agriculture and related natural resources; frequent articles on rangeland resources.

WATER RESOURCES RESEARCH
Publisher: American Geophysical Union, 1909 K St., N.W., Washington, D.C. 20006.
Publication: began 1965; present volume (1979), Vol. 15; issued bimonthly.
Evaluation: emphasizes social and natural sciences of water, including hydrology and watershed management.

WESTERN ASSOCIATION OF STATE GAME AND FISH COMMISSIONERS PROCEEDINGS
Publisher: host state game and fish department.
Publication: began 1921; present volume (1979), Vol. 59; issued annually; cumulative index, 1940-1969.
Evaluation: papers on wildlife management, programs, and law enforcement.

WILDLIFE SOCIETY BULLETIN (replaced Wildlife Society News)
Publisher: The Wildlife Society, Inc., 7101 Wisconsin Avenue, N.W., Washington, D.C. 20014.
Publication: began 1973; present volume (1979), Vol. 7; issued quarterly.
Evaluation: semitechnical articles on management, research, administration, law enforcement, education, and philosophy related to wildlife resources; book reviews and current literature section.

GLOSSARIES OF RANGE SCIENCE AND RELATED FIELDS [28]

Allaby, Michael. 1977. A Dictionary of the Environment. The Macmillan Press, London, Eng. 532 p.

Compiled in England; coverage is the environmental sciences generally, including ecology in its broadest sense.

American Society of Range Management, Range Term Glossary Comm. (Donald L. Huss, Chm.). 1964. A Glossary of Terms Used in Range Management. Amer. Soc. Range Mgt., Portland, Ore. 32 p.

A major attempt to bring together the language of range management; includes about 470 terms; purpose was to (1) assign and develop precise meanings to terms being used in range management and (2) create a dictionary on range management useful to all persons, both laymen and technicians; replaced by revised edition in 1974.

Canada Department of Agriculture, Research Branch. 1976 (Rev.). Glossary of Terms in Soil Science. Can. Dept. Agric. Pub. 1459. 44 p.

Technical definitions prepared by Nomenclature Committee of the Canadian Society of Soil Sciences; 8 1/2 x 11 in.; preliminary edition published in 1967.

Carpenter, J. Richard. 1956. An Ecological Glossary. Hafner Pub., Co., New York. 305 p.

Purpose: bring together the more technical and restricted usage of terms in ecology; author of Lincoln College, Oxford, England; originally published in 1938 by the University of Oklahoma Press; reprinted without revision in 1956.

Dayton, W.A. 1950 (Rev.). Glossary of Botanical Terms Commonly Used in Range Research. USDA Misc. Pub. 110. 41 p.

Includes primarily morphological and taxonomic terms with a minimum of ecological, physiological, and other botanical terms; first published in 1931, revised in 1950 and 17 new terms added; author a taxonomist with the U.S. Forest Service.

Ford-Robertson, F.C. (ed.). 1971. Terminology of Forest Science, Technology, Practice, and Products. Soc. Amer. For., Washington, D.C. 349 p.

Authorized by a Joint FAO/IUFRO Committee on Forestry Bibliography and Terminology; incorporates both British Commonwealth and U.S. usage; replaced FORESTRY TERMINOLOGY (1958, 3rd. Ed., SAF); gives substantial consideration to range science, outdoor recreation, and wildlife management.

Gray, Peter. 1967. The Dictionary of the Biological Sciences. Reinhold Pub. Corp., New York. 602 p.

Comprehensive coverage of biological terms emphasizing a "description of organisms, their anatomy, function, and mutual interactions"; author a professor of biology, University of Pittsburgh.

Hanson, Herbert C. 1962. Dictionary of Ecology. Philsosophical Library, New York. 382 p.

Emphasis given to terminology in applied ecology; includes many words from fields closely related to ecology such as forestry, range management, agronomy, soils, and genetics; author an ecology professor at The Catholic University of America.

Harris, Lorin E. 1966 (Rev.). Biological Energy Interrelationships and Glossary of Energy Terms. Natl. Acad. Sci.-Natl. Res. Council (Washington, D.C.) Pub. 1411. 35 p.

Sponsored by the Committee on Animal Nutrition, Nat. Res. Council; first published in 1962 under the title of GLOSSARY OF ENERGY TERMS.

Hutchinson, D.E. (Chm., SCSA Glossary Comm.). 1976. Resource Conservation Glossary. Soil Conserv. Soc. Amer., Arkeny, Iowa. 63 p.

Includes over 2,700 terms; intended to serve professionals, laymen, and students; covers 18 technologies: agronomy, biology, conservation, ecology, economics, engineering, forestry, fish and wildlife biology, geology, hydrology, mining, planning, pollution control, range science, recreation, soils, waste management, and water resources; originally published in 1952.

Ibrahim, Kamal. 1975. Glossary of Terms Used in Pasture and Range Survey Research, Ecology, and Management. FAO, Rome. 153 p.

Semitechnical terminology of soil-plant-animal interrelationships and the economic aspects of managing natural forage resources; intended for use by people with varying degrees of training in range management; author a range management specialist with FAO.

Lapedes, Daniel N. (Ed. in Chief). 1976. McGraw-Hill Dictionary of the Life Sciences. McGraw-Hill Book Co., New York. 907 + 38 p.

Emphasizes the basic life sciences including zoology, microbiology, genetics, anatomy, ecology, and general biology.

Meyer, Arthur, B., and F.H. Eyre (Eds.). 1958 (3rd Ed.). Forestry Terminology: A Glossary of Technical Terms Used in Forestry. Soc. Amer. For., Washington, D.C. 97 p.

Emphasizes terms used by foresters in their daily tasks; preapred by SAF Committee on Forestry Terminology and contributed to by subcommittee on range management and on recreation and wildlife; first edition published in 1944 and second edition in 1950; replaced by TERMINOLOGY OF FOREST SCIENCE, TECHNOLOGY, PRACTICE, AND PRODUCTS.

Schwarz, Charles F., Edward C. Thor, and Gary H. Elsner. 1976. Wildland Planning Glossary. USDA, For. Serv. Gen. Tech. Rep. PSW-13. 252 p.

Terminology principally relating to wildland planning, land utilization, range management, and forest management; includes more than 1400 terms considered most useful in wildland and related resource planning; prepared by the Pacific Southwest Forest and Range Experiment Station.

Seiden, Rudolph. 1957. Handbook of Feedstuffs--Production, Formulation, Medication. Springer Pub. Co., New York. 591 p.

Emphasis given to economic plants and other feedstuffs produced on ranges, farms, and factories; plant diseases, insecticides, herbicides, drugs; and general agricultural, botanical, chemical, and nutritional terms; definitions frequently expanded into encyclopedic explanations.

Society for Range Management, Range Term Glossary Comm. (M.M. Kothmann, Chm.). 1974 (2nd Ed.). A Glossary of Terms Used in Range Management. Soc. Range Mgt., Denver, Colo. 36 p.

Emphasis given to terms with unique or special usage in range management; first edition published in 1964; second edition includes 565 terms and incorporates extensive revision, particularly in terms related to grazing management; revision prepared by a special SRM committee.

Soil Science Society of America. 1973 (Rev.). Glossary of Soil Science Terms. SSSA, 677 South Segoe Road, Madison, Wisconsin. 33 p.

Prepared by SSA Committee on Terminology and covers the broad field of soil science; includes appendices prepared by special subcommittees including soil classification; first published in the 1956 SSSA Proceedings.

Swartz, Delbert. 1971. Collegiate Dictionary of Botany. The Ronald Press Co., New York. 520 p.

Covers the broad field of technical botany; contains nearly 24,000 entries.

Winburne, John N. (Ed.). 1962. A Dictionary of Agricultural and Allied Terminology. Mich. State Univ. Press, East Lansing. 905 p.

Comprehensive coverage of the wide scope of agricultural production, utilization, and agribusiness; includes range management terminology.

PUBLISHED BIBLIOGRAPHIES FOR RANGE SCIENCE [29]

Aldon, Earl F., and H.W. Springfield. 1973. The Southwestern Pinyon-Juniper Ecosystem: A Bibliography. USDA, For. Serv. Gen. Tech. Rep. RM-4. 20 p.

Entries classified by subject matter: ecological investigations, silvics, product utilization, range characteristics and wildlife values, water yield and sediment, and insects and diseases; not annotated; 457 references.

Anderson, E. William, and Robert W. Harris. 1973. References on Grazing Resources of the Pacific Northwest, 1896 through 1970. SRM, Pacific Northwest Section. 103 p.

Range ecosystems of Oregon, Washington, and British Columbia from 1896 through 1970 and their management; not annotated; 515 references.

Arthur, Louise M., and Ron S. Boster. 1976. Measuring Scenic Beauty: A Selected Annotated Bibliography. USDA, For. Serv. Gen. Tech. Rep. RM-25. 34 p.

Emphasizes the evaluation of scenic beauty of wildlands; annotated; many critical comments added; author index; 167 references.

Atkinson, J.J. 1971. A Bibliography of Canadian Soil Science. Can. Dept. Agric. Pub. 1452. 303 p.

Includes papers published through 1969; entries arrange chronologically by year and then alphabetically by author; author index and subject index; not annotated; 3,444 references.

Basile, Joseph V. 1967. An Annotated Bibliography of Bitterbrush (Purshia tridentata [Pursh] DC.). USDA, For. Serv. Res. Paper INT-44. 27 p.

Pertains primarily to restoring bitterbrush on depleted rangeland through revegetation and management; annotated; 221 references.

Blake, S.F. 1954. Guide to Popular Floras of the United States and Alaska. USDA Bibl. Bul. 23. 56 p.

Selected list of nontechnical works for the identification of flowers, ferns, and trees; annotated; arranged by regions and states; author index; 358 references.

Bovey, R.W., and J.D. Diaz-Colon. 1977. Selected Bibliography of the Phenoxy Herbicides. II. The Substituted Dibenzo-p-Dioxins. Texas Agric. Expt. Sta. Misc. Pub. 1323. 57 p.

Reviews TCDD, a toxic chlorodioxin sometimes formed in small amounts during the manufacture of 2,4,5-T; not annotated; 535 references.

_____. 1978. Selected Bibliography of the Phenoxy Herbicides. IV. Ecological Effects. Texas Agric. Expt. Sta. Misc. Pub. 1360. 28 p.

Alphabetically arranged by author; subject index; not annotated; 196 references; not annotated.

_____. 1978. Selected Bibliography of the Phenoxy Herbicides. VIII. Effects on Higher Plants. Texas Agric. Expt. Sta. Misc. Pub. 1388. 60 p.

Alphabetically arranged by senior author; subject index divided into research papers and review sections; covers effects including mode of action and manifested plant responses; 526 references; not annotated.

Campbell, R.S., L.K. Halls, and H.P. Morgan. 1963. Selected Bibliography on Southern Range Management. USDA, For. Serv. Res. Paper SO-2. 62 p.

Covers southern ranges (eastern Texas to the Atlantic, southern Missouri to the Gulf Coast), the domestic livestock and wildlife produced thereon, and the management of these lands, livestock, and wildlife; arranged by subject; not annotated; author index; about 1,200 references.

Christensen, Earl M. 1967. Bibliography of Utah Botany and Wildland Conservation. Brigham Young University Sci. Bul., Biol. Ser. 9 (1):1-136.

Includes references on range management and related natural resources management through 1964; arranged alphabetically by author; subject index; chronological list of authors; not annotated; over 2,560 references.

_____. 1967. Bibliography of Utah Botany and Wildland Conservation, No. II. Proc. Utah Acad. Sci. 44 (2):545-66.

A 1965-66 supplement to the previous reference; arranged alphabetically by author; 286 references.

Cushwa, Charles T. 1968. Fire: A Summary of Literature in the United States from the Mid-1920's to 1966. USDA, Southern For. Expt. Sta., Asheville, N. Car. 117 p.

Alphabetically arranged by author; entries arranged by subject matter in a second section; not annotated; 823 references.

Czapowskyj, Miroslaw M. 1976. Annotated Bibliography on the Ecology and Reclamation of Drastically Disturbed Areas. USDA, For. Serv. Gen. Tech. Rep. NE-21. 98 p.

Pertains mainly to mining effects and reclamation in the coal regions of the United States; annotated; arranged alphabetically by authors; includes material, subject, area, and multiple authorship indexes; 591 references.

Dalsted, Norman L., and F. Larry Leistritz. 1974. A Selected Bibliography on Coal-energy Development of Particular Interest to Western States. N. Dak. Agric. Econ. Misc. Rep. 16. 82 p.

Economic, sociological, and environmental impact of surface coal mining and reclamation; arranged by broad subject categories; author index; partly annotated; 486 references.

Diaz-Colon, J.D., and R.W. Bovey. 1976. Selected Bibliography of the Phenoxy Herbicides. I. Fate in the Environment. Texas Agric. Expt. Sta. Misc. Pub. 1303. 61 p.

Alphabetically arranged by author; author index; not annotated; 563 references.

_____. 1977. Selected Bibliography of the Phenoxy Herbicides. III. Toxicological Studies in Animals. Texas Agric. Expt. Sta. Misc. Pub. 1343. 105 p.

Alphabetically arranged by author; subject index; not annotated; 871 references.

_____. 1978. Selected Bibliography of the Phenoxy Herbicides. V. Interrelations with Microorganisms. Texas. Agric. Expt. Sta. Misc. Pub. 1379. 88 p.

Behavior of microorganisms in different ecological ecosystems when phenoxy herbicides are present; part A, herbicide effects on microorganisms; part B, effects of microorganisms on herbicides; alphabetically arranged by author; subject matter indexes; 308 references in part A, 202 in part B; not annotated.

Doell, Dean D., and Arthur D. Smith. 1965. A Selected Bibliography of Literature Applicable to Big Game Range Research. Utah State Div. Fish and Game, Salt Lake City. 93 p.

Covers big game biology and big game range management with western U.S. emphasis, about 1936 to 1964; alphabetically arranged by author; subject index; 1,398 references.

Dolnick, E.H., R.L. Medford, and R.J. Schied. 1976. Bibliography on the Control and Management of the Coyote and Related Canids With Selected References on Animal Physiology, Behavior, Control Methods, and Reproduction. USDA, Agric. Res. Serv., Beltsville, Md. 248 p.

Dorrance, M.J. 1966. A Literature Review on Behavior of Mule Deer. Colo. Dept. Game, Fish and Parks Spec. Rep. 7. 26 p.

Duvall, V.L., A.W. Johnson, and L.L. Yarlett. 1968. Selected Bibliography on Southern Range Management, 1962-1967. USDA, For. Serv. Res. Paper SO-38. 34 p.

A continuation of Cambell et al. (1963) through 1967; arranged by subject matter; about 650 references.

Eschmeyer, Paul H., and Van T. Harris (Eds.). 1974. Bibliography of Research Publications of the U.S. Bureau of Sports Fisheries and Wildlife, 1928-72. USDI, Bur. Sport Fish. and Wildl. Resource Pub. 120. 154 p.

Arranged alphabetically by research centers, including six wildlife stations; not annotated; no indexes.

Ffolliott, Peter F., and Warren P. Clary. 1972. A Selected and Annotated Bibliography of Understory-overstory Vegetation Relationships. Ariz. Agric. Expt. Sta. Tech. Bul. 198. 33 p.

Interactions in production, density, and composition of understories resulting from natural and modified overstories; annotated; arranged by overstory vegetation categories; author index; 262 references.

Frawley, Margaret L. (Comp.). 1971. Surface Mined Areas: Control and Reclamation of Environmental Damage, A Bibliography. USDI, Office Libr. Serv. Bibliog. Ser. 27. 63 p.

Giefer, G.J. 1976. Sources of Information in Water Resources, An Annotated Guide to Printed Materials. Water Info. Center, Inc., Point Washington, N.Y. 290 p.

Arranged by water resources subject categories; annotated; author-subject index; over 1,100 references.

Gifford, Gerald F., Don D. Dwyer, and Brien E. Norton. 1972. A Bibliography of Literature Pertinent to Mining Reclamation in Arid and Semi-arid Environments. Utah State Univ., Environment and Man Program. 23 p.

Arranged by subject categories; not annotated; 312 references.

Gray, James R. (Coord.). 1969. Economic Research in the Use and Development of Range Resources. Report No. 11. Range and Ranch Economics Bibliography. Western Agric. Econ. Res. Council, Las Cruces, New Mex. 199 p.

Arranged alphabetically by author in first section and by major subject categories in second section; no indexes; not annotated; over 1,000 references.

Heerwagen, Arnold J. 1971. A Selected Bibliography of Natural Plant Communities in 11 Midwestern States. USDA Misc. Pub. 1205. 30 p.

The nature of past and present plant communities and the response of these communities to environmental factors; alphabetically arranged by author; not annotated; 742 references.

Hibbs, L. Dale. 1966. Literature Review on Mountain Goat Ecology. Colo. Dept. Game, Fish, and Parks Spec. Rep. 8. 23 p.

Hopkins, Harold H. 1955. Literature of the Vegetation of Kansas. Trans. Kan. Acad. Sci. 58 (2):171-195.

Ecology, taxonomy, geography, range management, and soil conservation of Kansas vegetation; alphabetically by author; annotated; county index and chronological index; 339 references.

Horton, Jerome S. 1973. Evapotranspiration and Watershed Research as Related to Riparian and Phreatophyte Management: An Abstract Bibliography. USDA Misc. Pub. 1234. 192 p.

Relationships of vegetation to water loss, and effects on water yield of manipulating vegetation; references classified by subject; author index; annotated and abstracted; 713 references.

Horton, L.E. 1975. An Abstract Bibliography of Gambel Oak. USDA, For. Serv., Intermtn. Reg., Ogden, Utah. 64 p.

Ecology, taxonomy, and management of Gambel oak (*Quercus gambelii*); alphabetically by author; abstracted; subject index and author index; 123 references.

Hosley, N.W. 1968. Selected References on Management of White-tailed Deer, 1910 to 1966. USDI, Fish and Wildl. Serv. Sci. Rep. Wildl. 112. 46 p.

Arranged by subject categories; not annotated; about 700 references.

Kirsch, John B., and Kenneth R. Greer. 1968. Bibliography--Wapita-American Elk and European Red Deer. Mon. Fish. and Game Dept. Spec. Rep. 2. 147 p.

Klumph, S.G., and A. Haberer. 1975. Range: A Selected Bibliography of Research and Information. Alberta Dept. Energy and Natural Resources, Edmonton, Alberta. 201 p.

Literature of range management in Alberta and related areas through 1974; classified by subject; not annotated; over 1,300 references.

Krumpe, Paul F. 1976. The World Remote Sensing Bibliographic Index. Tensor Industries, Inc., Merrifield, Va. 600 p.

Leedy, D.L., T.M. Franklin, and E.C. Hekimian. 1976. Highway-Wildlife Relationships. Volume 2. An Annotated Bibliography. Natl. Tech. Info. Serv., Springfield, Va. 417 p.

Little, Elbert L., Jr., and Barbara H. Honkala. 1976. Trees and Shrubs of the United States--A Bibliography for Identification. USDA Misc. Pub. 1336. 56 p.

> Covers wild and cultivated woody plants of the 50 states, Puerto Rico, Virgin Islands, and Guam; 1950 to 1975, with some older publications; author index; not annotated; 470 references.

McGlinchy, S.E., R.A. Monson, and P. Nash. 1971. An Annotated Bibliography of the Wild Sheep of North America. Rachelwood Wildl. Res. Preserve (New Florence, Pa.) Pub. 1. 86 p.

Miller, G. 1977. The American Bison (*Bison bison*): An Initial Bibliography. Can. Wildl. Serv., Edmonton. 89 p.

Morris, Meredith J. 1967. An Abstract Bibliography of Statistical Methods in Grassland Research. USDA Misc. Pub. 1030. 222 p.

> Literature of the world through 1963; arranged alphabetically by author within subject matter categories; abstracted; author index; 1,118 references.

Neil, P.H., R.W. Hoffman, and R.B. Gill. 1975. Effects of Harassment on Wild Animals--An Annotated Bibliography of Selected References. Colo. Div. Wildl. Spec. Rep. 37. 21 p.

Patton, David R., and Peter F. Ffolliott. 1975. Selected Bibliography of Wildlife and Habitats for the Southwest. USDA, For. Serv. Gen. Tech. Rep. RM-16. 39 p.

> Research and management of important wildlife species and habitats in Arizona and New Mexico, 1913 to 1975; alphabetical by author; subject index; 390 references.

Pearson, H.A., C.E. Lewis, G.E. Probasco, and G.L. Wolters. 1973. Selected Bibliography on Southern Range Management, 1968-1972. USDA, For. Serv. Tech. Rep. SO-3. 50 p.

> A continuation of Duvall et al. (1968) and the third bibliography in a series; arranged by subject matter categories; not annotated; author index; about 950 references.

Peterson, Howard B., and Ralph Monk. 1967. Vegetation and Metal Toxicity in Relation to Mine and Mill Wastes: An Annotated Bibliography. Utah Agric. Expt. Sta. Cir. 148. 75 p.

> Alphabetically arranged by author; annotated-abstracted; author index; 167 references.

Renner, F.G., Edward C. Crafts, Theo. C. Hartman, and Lincoln Ellison.

1938. A Selected Bibliography on Management of Western Ranges, Livestock, and Wildlife. USDA Misc. Pub. 281. 468 p.

Covers western United States and adjacent Canada through 1933; with some references 1934-37; arranged by subject categories; not annotated; author index; 8,274 references.

Roubicek, C.B., et al. 1956-1957. Range Cattle Production--A Literature Review. 1. Reproduction; 2. Prenatal Development; 3. Birth to Weaning; 4. Post-weaning Performance; 6. Maternal Factors; 8. Effects of Climatic Environment. Ariz. Agric. Expt. Sta. Rep. 133, 135, 136, 138, 146, 154; 250 p.

Emphasis given to important aspects of range cattle production in western U.S.; combines review of literature with literature citation; coauthors vary: R.T. Clark, P.O. Stratton, O.F. Pahnish, and R.M. Richard; 1,211 citations.

Schmutz, Ervin M. 1978. Classified Bibliography of Native Plants of Arizona. Univ. Ariz. Press, Tucson. 160 p.

Contains over 3,000 references arranged in 30 subject matter categories; entries cover plant species and plant communities in Arizona including effects of management treatments.

Schultz, V., L.C. Eberhardt, J.M. Thomas, and M.I. Cochran. 1976. A Bibliography of Quantitative Ecology. Dowden, Hutchinson, and Ross, Stroudsburg, Pa. 384 p.

Schuster, Joseph L. (Ed.). 1969. Literature on the Mesquite (Prosopis L.) of North America. Texas Tech. Univ., ICASALS Spec. Rep. 26. 84 p.

Literature reviews on taxonomy and phytogeography, use, and natural enemies of mesquite, p. 4-24; bibliography section, p. 25-79, arranged alphabetically by author, containing many abstracts and annotations for the 374 references.

Squires, Victor R., and Henry F. Mayland. 1973. Selected Bibliography on Water-animal Relations. Water-Animal Relations Symp. Proc., USDA, ARS, Kimberly, Idaho, p. 209-239.

Emphasis given to water supplies, water requirements, animal performance, and water and livestock management; not annotated; 340 references.

Steel, Ordell, and William A. Berg. 1975. Bibliography Pertinent to Disturbance and Rehabilitation of Alpine and Subalpine Lands in the Southern Rocky Mountains. Environmental Resources Center, Colo. State Univ., Fort Collins, 104 p.

USDA, Soil Cons. Serv. 1977 (Rev.). List of Published Soil Surveys. USDA, Soil Cons. Serv., Washington, D.C. 13 p.

Soil surveys published by USDA based on work beginning in 1899; survey reports listed by state and then by county or region within state; date or dates of publication and availability for distribution indicated.

Vallentine, John F. 1978. U.S.-Canadian Range Management, 1935-1977: A Selected bibliography on Ranges, Pastures, Wildlife, Livestock, and Ranching. Oryx Press, Phoenix, Arix. 337 + 17 p.

Arranged by major and minor subject categories; covers English language literature pertaining to North America on range science, about 1935 to 1977; not annotated; author index; about 20,600 references.

West, Neil E. 1968. Ecology and Management of Salt Desert Shrub Ranges: A Bibliography. Utah Agric. Expt. Sta. Mimeo. Ser. 505. 30 p.

Literature published through August, 1968; arranged by author; not annotated; 367 references.

West, Neil E., Gerald F. Gifford, and Donald R. Cain. 1973. Biology, Ecology, and Renewable Resource Management of the Pigmy Conifer Woodlands of Western North America: A Bibliography. Utah Agric. Expt. Sta. Res. Rep. 12. 36 p.

Alphabetically arranged by author; not annotated; 828 references.

Zarn, M., T. Heller, and K. Collins (Comp.). 1977. Wild, Free-roaming Horses: An Annotated Bibliography, USDI, Bur. Land Mgt. Tech. Note 295. 54 p.

Chapter 4
AGENCIES AND ORGANIZATIONS AS INFORMATION SOURCES

FEDERAL AGENCIES

The purpose of the following discussion is to provide U.S. government sources of information that might be useful to individuals interested in range science and to give some idea of how more specific information can be obtained to meet individual needs.[30] Federal Information Centers, located in more than 30 states throughout the United States and the District of Columbia, may be contacted for information on various aspects of the federal government. Their exact location, address, and toll-free telephone numbers are given in the U.S. Government Manual, published by the U.S. Superintendent of Documents, U.S. Govt. Printing Office, Washington, D.C. 20402.

U.S. Department of Agriculture

On May 15, 1862, the U.S. Department of Agriculture was created by Congress as the 8th Executive Department of the federal government. Almost all Americans are served each day the the United States Department of Agriculture (USDA).[31] The principal task of this department is to improve farm income and develop domestic and foreign markets for agricultural products. Food programs are an important part of this agency's work to help alleviate poverty, hunger, and malnutrition among people in the United States and throughout the world. Protection of soil, water, forest, and other natural resources is an important task along with rural development and conservation programs and regulatory responsibilities as dictated by national policy. USDA also has a research function that directly or indirectly benefits all Americans.[32, 33]

Included within the reorganization of USDA effective January 24, 1978, the Science and Education Administration (SEA) was created. This new agency reflects the consolidation of the former Agricultural Research Service (now SEA-Agricultural Research), the former Cooperative State Research Service (now SEA-Cooperative Research), the Extension Service (now SEA-Extension), and the former National Agricultural Library (now SEA-Technical Information Systems). Also elsewhere within USDA the Economic Research Service was combined with the Farmer Cooperative Service and the Statistical Reporting Service to form the Economics, Statistics, and Cooperatives Service. This reorganization left unchanged the U.S. Forest

Forest Service, the Soil Conservation Service, the Agricultural Stabilization and Conservation Service, and the Agricultural Marketing Service as distinct agencies.

General USDA sources of information can be obtained for the following areas: consumer activities, environment, publications, speakers, and films. This information may be obtained from the Office of Communications, U.S. Department of Agriculture, Washington, D.C. 20250. Lists of publications are also available from this address.[34,35]

The following material lists the departments and agencies of government that have the most direct relationship to the field of range management although it is recognized that many others are important for specific purposes. The specific research and information and education agencies are discussed.

RESEARCH ORGANIZATIONS

The U.S. Department of Agriculture conducts significant range science and related research in three principal agencies. These agencies are Agricultural Research (SEA-AR), Forest Service (FS), and the Economics, Statistics, and Cooperatives Service (ESCS). Cooperative Research (SEA-CR) coordinates research activities between the federal government and the state agricultural experiment stations (SAES). The Technical Information Systems (formerly National Agricultural Library), while not being an active field research entity, is an integral part of the research program within the U.S. Department of Agriculture as well as providing valuable services to the agricultural research community.

SCIENCE AND EDUCATION ADMINISTRATION-AGRICULTURAL RESEARCH. SEA-AR was established on November 2, 1953, as the then Agricultural Research Service. Its principal mission is to provide essential knowledge and technology to increase the efficiency of production from the nation's agricultural resources and to conserve the natural resources for sustained production of food and fiber for an ever-growing population. This agency maintains close cooperation with other federal agencies, states, industry, foundations, and private groups. Agricultural Research has four regional offices, each representing twelve to fifteen states. The western, southern, and north central regions are particularly important sources of information on range management because their research programs include investigations on rangeland. The northeastern region, while the extent of rangeland is not as significant as in the others, is important because much of the research conducted there on forages is applicable to range management. The following addresses show the locations of the regional headquarters and the states for which those headquarters are responsible.

REGIONAL OFFICES OF SCIENCE AND EDUCATION ADMINISTRATION-AGRICULTURAL RESEARCH

Regions	Address
NORTHEASTERN: Maine, Connecticut, Massachusetts, Vermont, New Hampshire,	Bldg. 003, Agricultural Research Ctr., West Beltsville, Md. 20705

Rhode Island, New York, Pennsylvania, Delaware, Maryland, New Jersey, West Virginia	
NORTH CENTRAL: North Dakota, South Dakota, Nebraska, Kansas, Minnesota, Iowa, Missouri, Wisconsin, Illinois, Indiana, Michigan, Ohio, Alaska	2000 W. Pioneer Pkwy., Peoria, Ill. 61614
SOUTHERN: Kentucky, Virginia, Tennessee, North Carolina, Mississippi, Alabama, Georgia, South Carolina, Florida, Puerto Rico, Virgin Islands, Oklahoma, Texas, Arkansas, Louisiana	P.O. Box 53326, New Orleans, La. 70153
WESTERN: Washington, Oregon, California, Nevada, Idaho, Utah, Arizona, Montana, Wyoming, Colorado, New Mexico, Hawaii	2850 Telegraph Ave., Berkeley, Calif. 94705

Research locations throughout the United States can be quite helpful as sources of information and should be contacted by anyone interested in range management research information. The principal locations of SEA-AR range management research are shown in the following list.

NATIONAL PROGRAM SERVICE ADDRESSES, USDA, SEA-AR

Northern Great Plains Res. Lab.
P.O. Box 459
Mandon, N.D. 58554

Univ. of Nebraska
Rm. 316, Keim Hall
E. Campus
Lincoln, Neb. 68583

Veterinary Toxicology & Entomology Res. Lab.
P.O. Box Drawer GE
College Station, Tex. 77840

Univ. of Arizona
Tucson, Ariz. 85721

Rocky Mtn. Forest & Range Exp. Stat.
Flagstaff, Ariz. 86001

Northern Grain Insects Res. Lab
RR 3
Brookings, S.D. 57006

Bee Res. Lab
2000 East Allen Road
Tucson, Ariz. 85719

Soil Phosphorus Lab.
Rm. 7, Plant Sci. Bldg.
Colo. State Univ.
Ft. Collins, Colo. 80523

Crops Genetics & Improvement Res.
Dept. of Soil & Crops Sci.
Texas A&M Univ.
College Station, Tex. 77843

Beef Cattle Research
Southwestern Livestock & Forage Res. Sta.
Route 3
El Reno, Okla. 73036

Crops Research Lab.
Bay Road South of West Prospect
Ft. Collins, Colo. 85023

USDA High Plains Grasslands Res. Sta.
Route 1, Box 698
Cheyenne, Wyo. 82001

Poisonous Plant Res. Lab.
1150 East 14th N.
Utah State Univ.
Logan, Ut. 84322

Northern Plains Soils and
Water Res. Ctr.
P.O. Box 1109
Sidney, Mon. 59270

Snake River Cons. Res. Ctr.
Route 1, Box 186
Kimberly, Ida. 83341

Wash. State Univ.
Johnson Hall
Pullman, Wash. 99163

Brush Control Research
Dept. of Range Science
Texas A&M Univ.
College Station, Tex. 77843

U.S. Southern Great Plains Field Sta.
2000 18th Street
Woodward, Okla. 73801

Univ. of Wyoming
P.O. 3354 Univ. Sta.
Laramie, Wyo. 82071

Rangeland Insect Lab.
S. 11th Ave.
Montana State Univ.
Bozeman, Mon. 59715

Northwest Watershed Res. Ctr.
306 N. Fifth St.
P.O. Box 2700
Boise, Ida. 83701

Renewable Resource Ctr.
Univ. of Nevada
920 Valley Rd.
Reno, Nev. 89512

Squaw Butte Expt. Sta.
P.O. Box 833
Burns, Ore. 97720

Grassland-Forage Res. Ctr.
P.O. Box 748
Temple, Tex. 76501

Southwest Rangeland
Watershed Res. Ctr.
422 E. Seventh St.
Tucson, Ariz. 85705

Utah State Univ.
UMC 63 Crops Res. Lab.
Logan, Ut. 84322

U.S. Range Livestock Exp. Sta.
Route 1, Box 3
Miles City, Mon. 59301

U.S. Sheep Experiment Sta.
Dubois, Ida. 83423

Biological Control of Weeds Lab.
Univ. of Calif.
1050 San Pablo Avenue
Albany, Calif. 94706

U.S. FOREST SERVICE RESEARCH. The U.S. Forest Service was established in 1905, with the principal responsibility to administer the federal forest reserves. The national forests have been administered for the public good with increased emphasis on multiple use, sustained yield management since 1960. The principal objective of the Forest Service is to promote, protect, improve, and utilize national forest resources for the public good. An associated responsibility is to develop a firm scientific base for improvement of these resources for society's needs. To do this, eight experiment stations and the Forest Products Laboratory have been established and are presently conducting research, some of which is in the area of range management. The addresses of these stations are shown in the following table.

FOREST SERVICE EXPERIMENT STATIONS

Station	Address
Intermountain Forest and Range Exp. Sta.	507 25th St., Ogden, Utah 84401
North Central Forest Exp. Sta.	1992 Folwell Ave., St. Paul, Minn. 55108
Northeastern Forest Exp. Sta.	370 Reed Road, Broomall, Pa. 19008
Pacific Northwest Forest and Range Exp. Sta.	809 N.E. 6th Ave., (P.O. Box 3141), Portland, Ore. 97208
Pacific Southwest Forest and Range Exp. Sta.	1960 Addison St. (P.O. Box 245), Berkeley, Calif. 94701
Rocky Mountain Forest and Range Exp. Sta.	240 W. Prospect St., Ft. Collins, Colo. 80521
Southeastern Forest Exp. Sta.	P.O. Bldg. (P.O. Box 2570) Asheville, N.C. 28802
Southern Forest Exp. Sta.	T-10210, Postal Services Bldg., 701 Loyola Ave., New Orleans, La. 70113
Forest Products Laboratory	P.O. Box 5130, N. Walnut St., Madison, Wisc. 53705

The Forest Service conducts research on America's forests and related lands, often in cooperation with universities and other federal agencies. Range management research is an integral part of Forest Service activities along with tree genetics, plant nutrition, forest harvesting methods, fire control, development of improved processes for utilization of forest products, and environmental protection.

These experiment stations maintain mailing lists to provide interested persons with lists of available publications and reports. For further information, contact Information Office, Forest Service, USDA, Washington, D.C. 20250. Questions will be transferred to individuals within the agency most able to assist individuals desiring specific information. Requests for information are usually directed to the discipline level staff members experienced in various areas of national forest research and management. The regional offices also have information services. At these locations the information office will direct your question to the appropriate staff officer for reply.

ECONOMICS, STATISTICS, AND COOPERATIVES SERVICE (ESCS). The Economics Division (formerly the Economic Research Service) serves as an intelligence agency or research group for USDA, other federal agencies, farmers, agricultural industry, and the general public. Research data collected by this agency are analyzed and made available through research reports and economic outlook and situation reports for the major commodities, including beef cattle.

These programs deal with broad issues crossing commodity lines within the entire agricultural sector of the United States. Such examples of programs include consumer demand and analysis, agricultural finance, farm imputs, pricing policy and program analyses, farm economy projections, and determination of changes in the agricultural structure. This agency also focuses attention on the world-wide supply and demand of agricultural products with specific emphasis on the impact of U.S. policy on world agricultural trade.

The research effort centers on the use, conservation, development, and control of natural resources and their contribution to local, state, regional, and national economies. Also involved are analyses of the impact of environmental concerns on agricultural policy.

While the major emphasis is on the broad picture of agricultural economy in the U.S., specific programs deal with the development of statistical information and economic analyses of the production and marketing of farm commodities. Such programs are (1) evaluation of the organization and performance of major commodity subsections of the economy, (2) costs and returns to farms and marketers, (3) situation and outlook for various commodities, (4) commodity price projections, (5) price spreads for commodities, and (6) analysis of U.S. farm commodity programs. For further information, contact the Information Division of Economics, Statistics, and Cooperatives Service, USDA, in Washington, D.C. 20250. There is also information available at each of the land grant universities where cooperative work is ongoing between the Economics Division and the state university.

The Statistical Services function of ESCS prepares estimates and reports of production, supply, price, and other items associated with the functioning of the U.S. agricultural economy. Included in these reports are statistics on range, livestock, and related products germane to the rangeland industries. These estimates are prepared through sample surveys and dispensed in the form of periodic reports through 44 state-federal offices serving all states.

SCIENCE AND EDUCATION ADMINISTRATION-COOPERATIVE RESEARCH. SEA-CR (formerly Cooperative State Research Service) administers federal grant funds for research in the broad field of agriculture, including forestry and range management.[36] Funds are channeled through state agricultural experiment stations and other state institutions in all 50 states and the District of Columbia. These funds are based on legislation enacted in 1890 concerning land grant universities with special emphasis on agricultural experiment stations at these universities. Cooperative Research guides and coordinates research in cooperation with state agricultural experiment stations, forestry schools, federal research agencies, and other institutions.

Information may be obtained from each state institution as well as the U.S. Department of Agriculture. Information on cooperative state-federal agricultural and forest research may be obtained through the Science and Education Administration-Cooperative Research, U.S. Department of Agriculture, Washington, D.C. 20250.

Each western land grant university in a source of experimental results developed in cooperation with SEA-CR. The following list denotes the state agricultural experiment stations established under the Hatch Act of 1887 in the seventeen western states and Alaska:

University of Alaska
College, Alaska 99701

University of Arizona
Tucson, Ariz. 85721

University of California
Berkeley, Calif. 94720

Colorado State University
Fort Collins, Colo. 80523

University of Idaho
Moscow, Idaho 83843

Kansas State University
Manhattan, Kansas 66502

Montana State University
Bozeman, Mon. 59715

University of Nebraska
Lincoln, Neb. 68503

University of Nevada
Reno, Nev. 89557

New Mexico State University
Las Cruces, N. Mex. 88003

North Dakota State University
Fargo, N. Dak. 58102

Oklahoma State University
Stillwater, Okla. 74074

Oregon State University
Corvallis, Ore. 97331

South Dakota State University
Brookings, S. Dak. 57006

Texas A&M University
College Station, Texas 77843

Utah State University
Logan, Utah 84322

Washington State University
Pullman, Wash. 99163

University of Wyoming
Laramie, Wyo. 82070

The following nonland grant institutions participating in the Cooperative Forestry Research Program through the auspices of the McIntire-Stennis Act are also important sources of range information.

Northern Arizona University
Flagstaff, Ariz. 86001

Humboldt State College
Arcata, Calif. 95521

Forest and Conservation Experiment Station
School of Forestry
University of Montana
Missoula, Mont. 59801

Stephen F. Austin State University
Nacogdoches, Texas 75961

University of Washington
Seattle, Wash. 98105

TECHNICAL INFORMATION SYSTEMS (SEA-TIS) (formerly NATIONAL AGRICULTURAL LIBRARY. The TIS is not involved in actual research but does provide assistance to government agencies, particularly those within USDA, universities, research institutions, agricultural associations, industry, individual scientists, farmers, and the general public. Its resources are in excess of 1.5 million volumes, and it is one of the largest agricultural subject matter libraries in existence. Subject matter includes not only range management but also botany, chemistry, entomology, forestry, livestock, plant pathology, veterinary medicine, zoology, and general agriculture.

The library is located at 10301 Baltimore Blvd., Beltsville, Md. 20705. Its information is available through loans, photocopies, and reference services. For further information on the collections and reference services, contact the reference librarian. Numerous bibliographic data bases are stored in the catalog-indexing system (CAIN) discussed previously.

INFORMATION-EDUCATION AGENCIES AND ORGANIZATIONS

SCIENCE AND EDUCATION ADMINISTRATION-EXTENSION. SEA-E (formerly the Agricultural Extension Service) cooperates with the land grant universities and county governments throughout the nation. These three have shared jointly the financing, planning, and conduct of widespread agricultural education programs since establishment in 1914. Individuals interested in range management should contact the local county extension agent as the first source of information. This individual may be a range extension specialist if the county is predominately rangeland. More commonly, the county agent will be an expert in some other area and will refer you to the state extension specialist in range management who usually resides at the land grant university. Primary areas of assistance are agricultural production, marketing, home economics, nutrition, natural resources, 4-H youth development, rural development, and other agriculturally related subjects and programs. The national extension staff provides technical assistance to the state extension specialists and works directly with personnel of other federal agencies at the national level.

County extension agents work directly with individuals and groups to share present technology in solving everyday problems. Their activities also benefit agricultural businesses that serve farmers and ranchers. Further information on agricultural extension activities and information available may be obtained from the information services staff, SEA-Extension, Department of Agriculture, Washington, D.C. 20250. For local problems, contact the county extension agent at the county seat in your area.

WESTERN LIVESTOCK MARKETING INFORMATION AND OUTLOOK PROJECT. An important source of information to production agriculture related to rangelands in the seventeen western states is the Western Livestock Marketing Information and Outlook Project based in Denver, Colorado. This is a cooperative project among the federal level of SEA-Extension and the Economics Branch of ESCS in USDA and the state extension services in the 11 western and six plains states. Their primary purpose is to develop, analyze, and disseminate market information and outlook through three principal media: (1) the Western Livestock Round-Up which is disseminated through state livestock marketing specialists at the land grant universities in the 17 participating states, (2) weekly mailings of tables, charts, and analysis of pertinent market factors to extension economics specialists, and (3) response to individual requests from public and private sectors for market information and analytical reports.

The content of the Western Livestock Round-Up each month varies slightly. Principal components of the publication include demand analyses for livestock products, price spread, world meat trades, futures markets, feed grains, competitive products, sheep-lamb and wool market information, structural changes in marketing systems, and range and pasture conditions as dictated by precipitation, snow pack, and forecast of these conditions.

Users of the Western Livestock Round-Up and related material are producers, distributors, processors and consumers of livestock products, financial institutions, public employees, etc. The principal value to users is analysis of market trends and general market outlook information.

Those interested in further information should contact their local county extension agent, the extension economist at the land grant university or the Western Livestock Marketing Information Project, 2490 West 26th Avenue, Denver, Colo. 80211.

SOIL CONSERVATION SERVICE. Almost 3,000 local soil and water conservation districts, including some 2.3 million cooperating landowners and operators, carry out programs developed by the Soil Conservation Service (SCS). These national conservation programs include watershed projects, Great Plains Conservation Programs, resource conservation and development projects, recreation enterprises, and river basin studies. The SCS also prepares soil survey reports and snow survey reports.

While many SCS programs are directly beneficial to the management and understanding of rangelands, in recent years SCS services have been used by private individuals and by state, county, and local governments as sources of information and expertise to aid in developing, planning, and zoning requirements. Consequently, these and other non-farm groups are being assisted by this action agency.

The principal services provided by SCS in range resource management to non-USDA persons are technical assistance to nonfederal landowners in planning and applying conservation practices affecting soil, water, plants, and animals.

Particular publications useful in understanding SCS range programs and activities are: The National Range Handbook,[37] Agricultural Handbook 235 entitled "Classifying Rangeland for Conservation Planning,"[38] and career brochures outlining range conservationist opportunities in SCS.[39] The Soil Conservation Service has a technical staff of range conservationists along with soil conservationists and scientists, economists, agricultural and irrigation engineers, agronomists, biologists, foresters, plant materials specialists, geologists, landscape architects, and resource planning specialists. These provide technical staff assistance to individuals and groups on a wide range of programs that include range, pasture, cropland and woodland conservation, plant materials selection and availability, disturbed area reclamation, wildlife conservation, soil surveys, land inventory, and snow monitoring and surveys. Other activities include watershed and water resource activities in river basin investigations.

For those landowners who elect to develop SCS-guided resource conservation and development projects, SCS will provide technical assistance on these and other Great Plains conservation programs.[40] In doing this, there can be some cost sharing of conservation projects such as terracing, pond development, etc.

Range conservation programs by the SCS are geared to produce forage for livestock and maintain wildlife habitat and watershed and aesthetic values. SCS personnel can help individual landowners prepare and apply grazing use and management plans to maintain and improve the production and quality of the vegetation and to increase returns to land and management while maintaining a quality environment. These services include identification and inventory of land resources and evaluating their potential production, selection of management plans and treatment that insure optimum use of the resources, and selection of the appropriate conservation measures needed to assure proper conservation.

U.S. FOREST SERVICE. The Forest Service applies principles of multiple-use management for sustained yield on the nation's renewable forest resources. Management of almost 200 million acres in the National Forest System in cooperation with state and private forest industry as well as private and public agencies is an important aspect of this agency's work. Principal activities include timber production, livestock grazing, outdoor recreation, fish and wildlife habitat, watershed production, and aesthetics.

The U.S. Forest Service applies its management through a system of regions administered by regional foresters spread throughout the United States. The following lists the region and the address for each location within the United States.

REGIONAL HEADQUARTERS

NATIONAL FOREST SYSTEM REGIONS

Region		Address
1.	Northern	Federal Bldg., Missoula, Mon. 59807
2.	Rocky Mountain	11177 W. 8th Ave. (P.O. Box 25127), Lakewood, Colo. 80225
3.	Southwestern	Federal Bldg., 517 Gold Ave. SW., Albuquerque, N. Mex. 87102
4.	Intermountain	324 25th St., Ogden, Utah 84401
5.	California	630 Sansome St., San Francisco, Calif. 94111
6.	Pacific Northwest	319 SW Pine St. (P.O. Box 36231), Portland, Ore. 97208
7.	Southern	1720 Peachtree Rd., NW., Atlanta, Ga. 30309

8.	Eastern	633 W. Wisconsin Ave., Milwaukee, Wisc. 53203
9.	Alaska	Federal Office Bldg. (P.O. Box 1628), Juneau, Alaska 99802

Technical assistance is provided to state and private forestland owners and operators and processors of forestry products in the management of their operations to maximize efficiency of production and maintain conservation goals.

AGRICULTURAL STABILIZATION AND CONSERVATION SERVICE. The ASCS, established in June 1961, administers specific commodity and related land-use programs designed to adjust production with demand, protect natural resources, and stabilize farm income. These programs are administered through local farmer committees in each of the 2,700 agricultural counties in the United States.

ASCS deals in commodity programs and production adjustment efforts which are indirectly related to range management. Through the Agricultural Conservation Program (ACP), cost sharing, generally on a 50-50 basis, is provided to farmers and ranchers to carry out needed conservation and environmental measures, either under annual or long-term agreements. These programs, as discussed earlier, are designed by the Soil Conservation Service. The emphasis is on conservation and environmental practices that preserve the environment and sustain benefits to the public at least possible cost and to increase forest and rangeland production.

There is a forestry incentive program that cost-shares tree planting and timber improvement on private lands. This agency also provides emergency assistance in certain areas following certain widespread disasters. For example, Commodity Credit Corporation feed grains may be made available to livestock producers at reduced prices following drought conditions. Disaster payments can be made to farmers and ranchers participating in wheat, feed grain, and cotton programs to offset low income resulting from natural forces such as drought, hail damage, etc. For further information, contact the Information Division, Agricultural Stabilization and Conservation Service, U.S. Department of Agriculture, Washington, D.C. 20250. The first line of contact should be your local Soil Conservation Service, soil and water conservation district, or county agent.

U.S. Department of Interior

The Department of Interior (USDI) administers over 500 million acres of federal lands, much of which is rangeland located in the western United States. This department has been in existence for almost 130 years since March 3, 1849. One of the earliest tasks of this government agency was to administer the trust responsibilities for about 50 million acres of Indian reservations. This department's responsibilities focus on conservation, development, and utilization of a wide variety of reserves and resources, including mineral, water, land, and fish and wildlife resources. Coordination of federal and state recreation programs and the development of coal and oil shale resources are other major tasks. This department also has limited responsibility for research which is administered by the various agencies within USDI.[41]

Several agencies within the Department of Interior are important sources of information for range scientists, range managers, and others interested in this vast renewable resource. Inquiries should be directed to the specified offices within the Department of Interior, Washington, D.C. 20240. More specific information related to agencies within USDI is given below.

BUREAU OF LAND MANAGEMENT. The Bureau of Land Management (BLM) was established in 1946 and given the responsibility to manage the unreserved federal lands and their incumbent resources. These responsibilities include the total management of 450 million acres of National Resource Lands located primarily in the Far West and Alaska, with only scattered parcels throughout the other states. It also has the subsurface resource management for an additional 310 million acres where the federal government has mineral rights as well as the mineral resources on the submerged lands of the outer continental shelf.

While information may be obtained from the Office of Public Affairs, Bureau of Land Management, Department of the Interior, Washington, D.C. 20240, additional sources of information are in the state offices (the locations and areas of responsibility are shown below).

FIELD OFFICES--BUREAU OF LAND MANAGEMENT

State Offices	Area of Responsibility	Address
Alaska	Alaska	555 Cordova St., Pouch 7-512, Anchorage, Alaska 99510
Arizona	Arizona	2400 Valley Bank Center, Phoeniz, Ariz. 85073
California	California	Federal Bldg., Sacramento, Calif. 95825
Colorado	Colorado	Colorado State Bank Bldg., Denver, Colo. 80202
Eastern States	All states bordering on and east of the Mississippi River	7981 Easter Ave., Silver Spring, Md. 20910
Idaho	Idaho	Federal Bldg., Boise, Ida. 83724
Montana	Montana, North Dakota, South Dakota	Granite Tower Bldg., 222 N. 32nd St., Billings, Mon. 59101

Nevada	Nevada	Federal Bldg., Reno, Nev. 89502
New Mexico	New Mexico, Oklahoma	Federal Bldg., Santa Fe, N. Mex. 87501
Oregon	Oregon, Washington	729 N.E. Oregon St., Portland, Ore. 97208
Utah	Utah	University Club Bldg., 136 E. S. Temple St., Salt Lake City, Utah 84111
Wyoming	Wyoming, Kansas, Nebraska	Federal Bldg., Cheyenne, Wyo. 82001

OUTER CONTINENTAL SHELF (OCS) OFFICES

Alaska	Alaska, OCS	P.O. Box 1159, Anchorage, Alaska 99510
New York	New York OCS (north from Florida-Georgia State line)	Federal Bldg., Suite 32-120, 26 Federal Plaza, New York City, N.Y. 10007
Gulf of Mexico	Gulf of Mexico and Florida OCS	Hale Boggs Federal Bldg., 500 Camp St., New Orleans, La. 70130
Pacific	Pacific OCS (including Hawaii OCS)	300 N. Los Angeles, Los Angeles, Calif. 90012

SERVICE AND SUPPORT OFFICES

Denver Service Center	Federal Center, Bldg. 50, Denver, Colo. 80225
Boise Interagency Fire Center	3905 Vista Ave., Boise, Ida. 83705

U.S. FISH AND WILDLIFE SERVICE. The U.S. Fish and Wildlife Service (US-FWS) has national responsibility in wildlife protection. Currently the U.S. Fish and Wildlife Service is responsible for management of over 350 national wildlife refuges consisting of more than 30 million acres, about 100 national fish hatcheries, 35 fish and wildlife research stations and laboratories, and cooperative research units at 45 major universities across the country. In addition, a national network of wildlife law enforcement agents and wildlife enhancement biologists is at work for this agency.

This agency has extensive conservation, education, and public information programs comprised of conservation education talks, TV and radio presentations, development and distribution of leaflets and brochures, operation of visitors centers, self-guided nature trails, etc. The management is comprised of a headquarters office in Washington, D.C. and regional offices. The following is a list of regional offices of the U.S. Fish and Wildlife Service that occur in areas of the United States where rangelands are prominent.

REGIONAL OFFICES--UNITED STATES FISH AND WILDLIFE SERVICE

Region	Address
ATLANTA-Alabama, Arkansas, Florida, Georgia, Kentucky, Louisiana, Mississippi, North Carolina, Puerto Rico, South Carolina, Tennessee, Virgin Islands	17 Executive Park Dr. N.E., Atlanta, Ga. 30329
ALBUQUERQUE-Arizona, New Mexico, Oklahoma, Texas	P.O. Box 1306, Albuquerque, N. Mex. 87103
ANCHORAGE-Alaska	813 D. St., Anchorage, Alaska 99501
DENVER-Colorado, Iowa, Kansas, Missouri, Montana, Nebraska, North Dakota, South Dakota, Utah, Wyoming	P.O. Box 25486, Denver Federal Center, Denver, Colo. 80225
PORTLAND-California, Hawaii, Idaho, Nevada, Oregon, Washington	P.O. Box 3737, Portland, Ore. 97208

Environmental questions may be directed to the Assistant Director, Public Affairs, USFWS, Washington, D.C. 20240, or to one of the regional headquarters locations. Publications may be obtained from the Superintendent of Documents, Washington, D.C., Government Printing Office, 20402.

NATIONAL PARK SERVICE. National Park Service (NPS) has existed since 1916 and was established in the Department of Interior to administer an extensive system of national parks, historic sites, monuments, and recreational areas. These are administered for the enjoyment and education of the American people and to protect the natural environment as well as assist states, local governments, and citizen groups in the development of park areas. The Service Center, National Park Service, Denver, Colorado, provides professional services, including planning, architectural, engineering, and other such services for improving the national parks. Further information concerning NPS may be obtained by contacting the Assistant to the Director for Public Affairs, National Park Service, Department of the Interior, Washington, D.C. 20240. The regional offices are listed below:

REGIONAL OFFICES--NATIONAL PARK SERVICE

Region	Address
NORTH ATLANTIC--Maine, New Hampshire, Vermont, Massachusetts, Rhode Island, Connecticut, New York, New Jersey	150 Causeway St., Boston, Mass. 02114

MID-ATLANTIC--Pennsylvania, Maryland, West Virginia, Delaware, Virginia	143 S. 3d St., Philadelphia, Pa. 19106
SOUTHEAST--Alabama, Florida, Georgia Kentucky, Mississippi, North Carolina, South Carolina, Tennessee, Puerto Rico, Virgin Islands	1895 Phoenix Blvd., Atlanta, Ga. 30349
MIDWEST--Ohio, Indiana, Michigan, Wisconsin Illinois, Minnesota, Iowa, Missouri, Nebraska, Kansas	1709 Jackson St., Omaha, Neb. 68102
ROCKY MOUNTAIN--Montana, North Dakota, South Dakota, Wyoming, Utah, Colorado	P.O. Box 25287, Denver, Colo. 80225
SOUTHWEST--Arkansas, Louisiana, New Mexico, Oklahoma, Texas	Box 728, Santa Fe, N. Mex. 87501
WESTERN--Arizona, California, Nevada, Hawaii	450 Golden Gate Ave., San Francisco, Calif. 94102
PACIFIC NORTHWEST--Alaska, Idaho, Oregon, Washington	1424 4th Ave., Seattle, Wash. 98101
NATIONAL CAPITOL PARKS--Metropolitan area of Washington, D.C.	1100 Ohio Dr., S.W., Washington, D.C. 20242

BUREAU OF INDIAN AFFAIRS. The major thrust of the Bureau of Indian Affairs (BIA) is to encourage and train both Indian and Alaskan natives to handle their own affairs under a trust relationship with the Federal Government. This agency works with native Americans on some 50 million acres of land. They carry out their work with a corps of specialists within the agency that includes range conservationists. Further information on BIA programs and the management of these rangelands can obtained from the Office of Public Information, Bureau of Indian Affairs, U.S. Dept. of the Interior, 1951 Constitution Ave., N.W., Washington, D.C. 20245.

Environmental Protection Agency

No other agency in modern times has had more impact on the lives of every citizen than the Environmental Protection Agency (EPA). This agency, established on December 2, 1970, coordinates governmental action on preservation and conservation of the environment. Its principal goals are to control and abate air, water, solid waste, pesticide, noise, and radiation pollution by determination of the magnitude of various pollution problems through coordinated research and regulatory activities with local and state governments, industry, other federal agencies, private and public groups, individuals, and educational institutions. Of particular interest to range managers are the EPA regulations regarding the control of nonpoint source pollution. Other principal activities of EPA include providing policy guidelines for enforcement of regulations and research and development. EPA conducts a research program pursuant to development of technological controls of pollution. This is accomplished in EPA's national laboratories and through grants to outside researchers.

Through guidelines developed by EPA, all other federal agencies and private corporations interacting with these agencies are required to describe the impact of their operations on the environment and to publish environmental impact statements after public review and comment.

Regional offices located throughout the United States are the agency's principal representative for contacts with the public and other public agencies. The addresses for these offices are:

Region	Address
I	John F. Kennedy Federal Bldg., Boston, Mass. 02203
II	26 Federal Plaza, New York, N.Y. 10007
III	Curtis Bldg., 6th & Walnut St., Philadelphia, Pa. 19106
IV	1421 Peachtree St., N.E., Atlanta, Ga. 30309
V	230 S. Dearborn St., Chicago, Ill. 60604
VI	1600 Patterson St., Dallas, Texas 75201
VII	1735 Baltimore Ave., Kansas City, Mo. 64108
VIII	1860 Lincoln St., Denver, Colo. 80203
IX	100 California St., San Francisco, Calif. 94111
X	1200 6th Ave., Seattle, Wash. 98101

CANADIAN DEPARTMENT OF AGRICULTURE (RESEARCH BRANCH)

The program services division within the research branch is responsible for assembling and disseminating scientific and technical information on all phases of research conducted by this branch.[42] The research is organized into four broad research groups: crops, animals, production, and protection--including research more specifically on cereal crops, forage crops, beef cattle, sheep, environment and resources, food, environmental quality, entomology, plant pathology, and weeds. Many of these areas of research are closely related to range management and represent valuable information resources. Research is carried out at problem-oriented stations located throughout Canada. The following is a listing of some of the more important stations related to range management research.

KAMLOOPS, BRITISH COLUMBIA. This research station studies the problems of the ranching industry in British Columbia. The research program places emphasis on rangeland vegetation, dynamics and management, determination of carrying capacity for livestock, poisonous plants, and range seeding. Over 120,000 acres of rangeland are available for studies. Other pasture and forage crops production research is also conducted.

BEAVERLODGE, ALBERTA. The research work at this location includes development of proper land husbandry for pasture and hay production, management of tame and wild pollinators of legumes, and the management of livestock.

LETHBRIDGE, ALBERTA. Projects ongoing at this location include range management, forage plant breeding, beef and sheep breeding, and the problems of dryland farming. This location also serves as a source of direction for work at the Manyberries and Vauxhall substations located nearby.

MELFORT, SASKATCHEWAN. The principal objective at this station is centered on pasture improvement and forage utilization in the prairie provinces of Canada. The studies on forages, seed production, weed control, and beef cattle nutrition are also important functions there.

REGINA, SASKATCHEWAN. Forage crops studies and weeds research for the prairie provinces are conducted. These scientists deal with problems of cultural and chemical control of weeds on native rangelands with emphasis on determination of the fate of chemicals in plants and in soils.

SASKATOON, SASKATCHEWAN. At this station ecological research is conducted that is applicable to range management. The University of Saskatoon is the location of Canada's International Biological Program's Matador Project. The Saskatchewan unit of the National Soil Survey is also located at this station.

SWIFT CURRENT, SASKATCHEWAN. Research scientists at this station include a forage crop team specializing in development of drought-tolerant grasses and legumes and the management of such plants under dryland conditions.

OTTAWA, ONTARIO. Forage crops research and nutritional requirements and overwintering of beef cattle are studied at this station along with many other phases of agricultural research. Also at Ottawa is the Animal Research Institute that conducts studies on the improvement of the productive efficiency of Canadian livestock, including beef cattle and sheep. Also investigated are pure and mixed species of forages, the nutritional requirements of calves, milk cows, and sheep and the physiology of reproduction of sheep and cattle. Research is also ongoing on the metabolic diseases, including ketosis and white muscle disease in cattle.

Further information may be obtained on the research activities of the Canadian Department of Agriculture by contacting Information Services, Canada Department of Agriculture.[43,44] Information may also be obtained by contacting individual research stations. A current publications list[45] is made available for those interested, also a publication on the history of Canadian Agriculture.[46] Within the Information Division there are two divisions, new media services and public services. The latter section disseminates information to the public on an individual request basis.

PROFESSIONAL ORGANIZATIONS

Professional and technical societies and organizations represent memberships devoted to the study, development, and dissemination of knowledge of their respective disciplines. This section is designed to serve as an abbreviated guide to the associations and organizations that may have a particular value to individuals seeking additional information in range science and related fields. Such organizations can also supply information of current activities of professionals in research, management, and administration of natural resources. Such sources of information may lead to contact with highly qualified individuals that can provide insight to current events. This list is not exhaustive but includes those organizations and societies deemed to be good sources of information about range science and related fields.

AGRICULTURAL INSTITUTE OF CANADA (AIC) (formerly Canadian Society of Technical Agriculturalists to 1945)
Address: 151 Slater St., Suite 907, Ottawa, Ontario K1P 5H4

The Agricultural Institute of Canada was founded in 1920 for the purpose of promoting the science of agriculture and its related industries. This work is accomplished through the provincial branches and through various subject matter affiliates that include the animal, entomology, plant, and soil sciences.

Publications of the Agricultural Institute of Canada include the Agricultural Institute Review, Canadian Journal of Plant Science, Canadian Journal of Animal Science, and the Canadian Journal of Soil Science.

THE AMERICAN AGRICULTURAL ECONOMICS ASSOCIATION (AAEA) (formerly American Farm Management Association to 1918, American Farm Economics Association to 1967)
Address: Agricultural Experiment Station, Rm. 221, University of Kentucky, Lexington, Ky. 40506

AAEA was founded in 1910 to further the development of knowledge of agricultural economics for improving agriculture, and to further an understanding of agriculture's contribution to the general economy. Its membership includes state, federal, and industrial agricultural economists from the research, teaching, extension, and agribusiness sectors. Publications of the American Agricultural Economics Association include the American Journal of Agricultural Economics and the American Bibliography of Agricultural Economics (1971-74).

AMERICAN FORAGE AND GRASSLAND COUNCIL (AFGC) (formerly Joint Committee on Grassland Farming to 1958, American Grassland Council to 1966)
Address: 121 Dantzler Court, Lexington, Ky. 40503

The AFGC was founded in 1944 to provide stimulation for increasing the production and utilization of quality forage through grassland farming and to distribute new information on grassland farming to scientists and the general public.

Publications of the AFGC include the Forage and Grassland Progress (quarterly) and an annual report of the research-industry conference. Other publications include a membership directory (annually) and other books and booklets stressing grassland farming practices.

AMERICAN FORESTRY ASSOCIATION (AFA)
Address: 1319 18th St., N.W., Washington, D.C. 20036

AFA was founded in 1875 as a citizens' conservation organization working to advance the management and use of forests and related natural sources. Publications include the American Forest Magazine and the NATIONAL REGISTRY OF CHAMPION BIG TREES AND FAMOUS HISTORICAL TREES as well as other books and reprints.

AMERICAN INSTITUTE OF BIOLOGICAL SCIENCES (AIBS)
Address: 1401 Wilson Blvd., Arlington, Va. 22209

AIBS was founded on November 21, 1946, as a nonprofit scientific and educational association to further the advancement of the agricultural, biological, and medical sciences through research. The Institute assists societies, other organizations, and biologists and cooperates with local, national and international organizations to promote the application of biological sciences to the betterment of mankind. Publications of AIBS include the monthly journal Bioscience and the monthly AIBS Newsletter. Professional activities are numerous but include symposia, conferences and meetings such as the International Botanical Congress, placement services, and the National Register of Scientific and Technical Personnel.

AMERICAN SOCIETY OF AGRONOMY (ASA)
Address: 677 S. Segoe Road, Madison, Wisc. 53711

The American Society of Agronomy (ASA) was founded in 1907. Its membership is made up of agronomists, plant breeders, physiologists, soil scientists, chemists, educators, technicians, and others dedicated to improved crop production and soil management. Membership also includes university and college student members who are associated with the various agronomy clubs. Constituent societies include the Soil Science Society of America and the Crop Science Society of America.

Professional activities include annual conventions and regional meetings. Its publications include the Agronomy Journal, Journal of Agronomic Education, the Journal of Environmental Quality, Agronomy News, Agronomy Abstracts (annual), and Crops and Soils magazine. Other special publications include FORAGE FERTILIZATION (1974), FORAGE PLANT PHYSIOLOGY AND SOIL-RANGE RELATIONSHIPS (1964), RANGE RESOURCES OF THE SOUTHEASTERN STATES (1973), TROPICAL FORAGES IN LIVESTOCK PRODUCTION SYSTEMS (1975), FORAGE ECONOMICS-QUALITY (1968), and BIOLOGICAL NITROGEN FIXATION IN FORAGE LIVESTOCK SYSTEMS (1976).

AMERICAN SOCIETY OF ANIMAL SCIENCE (ASAS) (formerly American Society of Animal Nutrition to 1912, American Society of Animal Production to 1961)
Address: c/o David C. England, Department of Animal Science, Oregon State University, Corvallis, Oreg. 97331

ASAS was established in 1908 for the purpose of promoting the improvement of methods of investigation, instruction, and extension in animal production. Membership includes all persons engaged in or previously involved with the production of livestock products or allied businesses. Professional activities include annual meetings held each summer, regional meetings, and the publication of the Journal of Animal Science (monthly).

AMERICAN SOCIETY OF FARM MANAGERS AND RURAL APPRAISERS (ASFMRA) (formerly American Society of Farm Managers to 1936)
Address: 210 Clayton Street, Denver, Colo. 80206

The ASFMRA was founded in 1929 for the purpose of accrediting individuals as farm managers and rural appraisers and to promote the science and application of farm and ranch management and rural appraisal. Membership includes professional farm managers and appraisers as well as researchers, teachers, and extension workers in farm and ranch management and rural appraisal.

Publications include the ASFMRA Journal, ASFMRA Newsletter, PROFESSIONAL RURAL APPRAISAL MANUAL (1975) and the PROFESSIONAL FARM MANAGEMENT MANUAL (1972). This association also publishes periodic newsletters and holds annual conventions.

AMERICAN SOCIETY OF PLANT PHYSIOLOGISTS (ASPP)
Address: 9650 Rockville Pike, Bethesda, Md. 20014

ASPP membership consists of professional plant physiologists and plant biochemists engaged in research and teaching. Since 1924, this association has promoted research in plant physiology.

Meetings are held annually and programs and abstracts are published following the meeting. Publications include Plant Physiology (monthly), a bimonthly newsletter, and an annual directory.

AMERICAN SOCIETY OF PLANT TAXONOMISTS (ASPT)
Address: Department of Biology, University of Washington, Seattle, Wash. 98195

ASPT was founded in 1937 to promote the taxonomy and systematic botany of vascular and nonvascular plants including phylogeny, phytogeography, ecology, floristics, and similar subjects. Meetings are held annually, often in connection with the American Institute of Biological Sciences. Publications include Systematic Botany (quarterly) and a directory published at irregular intervals. ASPT also publishes an index of current taxonomic research.

AMERICAN VETERINARY MEDICAL ASSOCIATION (AVMA) (formerly U.S. Veterinary Medical Association to 1898)

Address: 930 N. Meacham Road, Schaumburg, Ill. 60196

AVMA was founded in 1863 and includes state and local groups as well as student chapters. It is a professional society of veterinarians and provides placement services and other services to its members. The objectives of AVMA are to foster the science and art of veterinary medicine including its relationship to agriculture and public health. Publications include the American Journal of Veterinary Research, the Journal of the American Veterinary Medicine Association, and a biennially published directory.

ARCHAEOLOGICAL INSTITUTE OF AMERICA (AIA)
Address: 53 Park Pl., Eighth Floor, New York, N.Y. 10007

This institute was founded in 1879 for the purpose of providing a means for the professional scholar to interact with interested laymen to keep abreast of the latest archaeological discoveries and their interpretation and significance. Its membership includes professional archaeologists as well as those individuals interested in the field of archaeology. Publications include Archaeology and the American Journal of Archaeology.

ASSOCIATION OF CONSULTING FORESTERS (ACF)
Address: Box 369, Yorktown, Va. 23690

The ACF was formed in 1948 for the purpose of raising the professional standards of consulting foresters and to improve the work being done by these individuals. It also serves as a forum for the exchange of information and expression of opinions of consulting foresters.

AUSTRALIAN RANGELAND SOCIETY (ARS)
Address: c/o CSIRO, Division of Land Resources, Wembley, Western Australia

The Australian Rangeland Society was founded in 1977 for the purpose of directing research to problem areas of range management, to bridge communication gaps among researchers, graziers, land managers, extension workers, educators and the general public, and to encourage awareness of land resources management problems and solutions. Membership consists primarily of researchers, extension workers, land managers, and ranchers. Publications include the Australian Rangeland Journal and the quarterly Arid Zone Newsletter.

BOTANICAL SOCIETY OF AMERICA (BSA)
Address: c/o C.C. Baskin, School of Biological Sciences, University of Kentucky, Lexington, Ky. 40506

The Botanical Society of America was formed in 1906 from a merger of the Botanical Society, Society for Plant Morphology and Physiology, and the American Mycological Society. Its purpose is to promote botanical science and educational advancements by conducting special research programs and sponsoring periodic summer institutes. Publications include the American Journal of Botany and Plant Science Bulletin. Also published irregularly are career bulletins and a guide to graduate study in botany for the United States and Canada. Meetings are held annually, usually in conjunction with AIBS.

BRITISH GRASSLAND SOCIETY (BGS)
Address: c/o Grassland Research Institute, Hurley, Maidenhead, Berkshire, Eng. SL6 5LR.

BGS was founded in 1945 to advance the methodology of producing and utilizing grass and forage crops and to advance the education and research into grassland management.

Membership includes research workers, advisers and local grassland societies. Publications include the Journal of the British Grassland Society, the name of which was changed in 1979 to Grass and Forage Science.

CANADIAN SOCIETY OF AGRONOMY (CSA)
Address: 151 Slater Street, Ottawa, Ontario K1P 5H4

CSA was formed in 1955 to bring about closer professional cooperation between agronomists throughout Canada and to provide the opportunity to report, exchange, and evaluate information pertinent to agronomy. This organization is an affiliate of the Agricultural Institute of Canada.

Its principal publication is the Canadian Journal of Plant Science, published for the society by the Agricultural Institute of Canada.

CANADIAN SOCIETY OF ANIMAL SCIENCE (CSAS) (formerly Canadian Society of Animal Production)
Address: 151 Slater Street, Ottawa, Ontario K1P 5H4

CSAS was founded in 1950 to provide a discussion forum for problems of common interest among individuals in the animal industries. This organization also provides avenues for improvement and coordination of research, extension, and teaching among animal scientists and to encourage the publication of scientific and other educational materials pertaining to the animal industry.

This organization is also an affiliate of the Agricultural Institute of Canada, which publishes the Canadian Journal of Animal Science for the CSAS.

CANADIAN SOCIETY OF SOIL SCIENCE (CSSS)
Address: 151 Slater Street, Ottawa, Ontario K1P 5H4

CSSS was founded in 1954 to foster all branches of soil science. This organization is an affiliate of the Agricultural Institute of Canada, which publishes the Canadian Journal of Soil Science for the CSSS.

COUNCIL FOR AGRICULTURAL SCIENCE AND TECHNOLOGY (CAST)
Address: 250 Memorial Union, Ames, Iowa 50011

The purpose of CAST, since its incorporation in May 1972, is to advance the understanding and use of agricultural science and technology in the public's interest. CAST serves as a resource group to which the public and government may turn for information on agricultural issues. It organizes task forces of agricultural scientists and technologists to assemble and interpret factual information on critical issues and disseminates this information in a usable form to the public, news media, and government.

CAST reports represent consensus statements on various subjects as drafted by task forces of imminent authorities representing the various disciplines relevant to each subject. These authorities may or may not be members of CAST or CAST member societies. Task force members are chosen on the basis of their knowledge concerning scientific subjects.

CAST is a consortium of more than 25 agricultural science societies. Other nonprofit agricultural and agriculture-related societies that do not qualify for society membership can be associate society members. Other organizations contributing to the support of CAST can become either supporting or sustaining members. Libraries and other information centers represent subscriber members. Individuals supporting the objectives of CAST and wanting to keep informed of its activities can become individual members upon payment of annual dues.

CROP SCIENCE SOCIETY OF AMERICA (CSSA)
Address: 677 S. Segoe Road, Madison, Wisc. 53711

CSSA was originally formed as a Crop Science Division of the American Society of Agronomy in 1955. Its purpose has been to advance research, extension, and teaching of all basic and applied phases of the crop sciences with particular emphasis on the improvement, culture, management, and utilization of field crops. CSSA publishes Crop Science and cooperates with ASA on other publications of joint interest. Membership includes plant breeders, plant physiologists, ecologists, crop production specialists, seed technologists, and others interested in the improvement and use of field crops. Meetings are held annually in conjunction with American Society of Agronomy and Soil Science Society of America.

THE ECOLOGICAL SOCIETY OF AMERICA (ESA)
Address: Library 3131, The Evergreen State College, Olympia, Wash. 98505

ESA, founded in 1915, promotes the study of individual organisms in relationship to their environment and as members of populations, communities, and ecosystems. It also serves to stimulate ecological research and the interchange of ecological concepts and to assist in the application of ecological principles to agriculture, forestry, wildlife management, range management, conservation of natural resources, and related industries. Its membership is comprised of educators, professional ecologists, and scientists in related fields interested in plant and animal ecology. It serves its membership and others by the publication of Ecology, Ecological Monographs, and the ESA Bulletin.

THE ENTOMOLOGICAL SOCIETY OF AMERICA (ESA)
Address: 4603 Calvert Road, College Park, Md. 20740

ESA is a professional society of entomologists and others interested in the study and control of insects. ESA was founded in 1953 for the purpose of promoting the science of entomology and all its branches and to assure cooperation among entomologists and associated scientists. Its membership includes most persons engaged in entomological work and others in allied fields of science. ESA publications include the Journal of Economic Entomology, Annals of the Entomological Society of America, the Bulletin of the Entomological Society of America, and Environmental Entomology.

NATIONAL AUDUBON SOCIETY (NAS)
Address: 950 Third Ave., New York, N.Y. 10022

The National Audubon Society was organized in 1901 and incorporated in 1905. Its purpose was to educate the public on the value and need for conservation of wild birds and other animals, plants, soil, and water resources. Publications include the Audubon Magazine and Audubon Field Notes.

NATIONAL WILDLIFE FEDERATION (NWF)
Address: 1412 16th Street, N.W., Washington, D.C. 20036

NWF was founded in 1936 and consists of a federation of 53 state and territorial conservation organizations plus associate and individual members. Its purpose is to encourage wise management of natural resources and to provide and promote a greater appreciation of these resources. This organization gives organizational and financial help to local conservation projects.

NWF publications include the weekly Conservation Report, Conservation News (semimonthly), the International Wildlife Magazine (bimonthly), National Wildlife Magazine (bimonthly), Conservation Directory (annually), and other news reports and magazines.

NATURE CONSERVANCY (NC)
Address: 1800 N. Kent Street, Suite 800, Arlington, Va. 22209

Nature Conservancy was founded in 1917 and is a nonprofit membership organization dedicated to the preservation of natural areas for present and future generations. It cooperates with colleges, universities, and public and private conservation organizations to obtain lands for scientific and educational purposes.

Its publications include the Nature Conservancy Preserve Directory plus other newsletters.

SOCIETY FOR RANGE MANAGEMENT (SRM) (American Society of Range Management until 1971)
Address: 2760 West Fifth Avenue, Denver, Colo. 80204

SRM was organized in 1948 to develop an understanding of range ecosystems and the principles applicable to the management of range resources; to assist all who work with range resources to keep abreast of new findings and techniques in the science and art of range management; to improve the effectiveness of range management; to obtain from range resources the products and values necessary for man's welfare; to create a public appreciation of the economic and social benefits to be obtained from the range environment; and to promote professional development of its members. Membership, which includes about 5,500 individuals, representing some 50 countries, consists of ranchers, research scientists, government agency administrators, technical assistance personnel, educators, students, agribusiness people, and others interested in the management of range resources for livestock, wildlife, watershed, recreation, and environmental quality.

Publications include the Journal of Range Management, Rangelands (formerly Rangeman's Journal), Rangeman's News (1969-1974 only), A GLOSSARY OF TERMS USED IN RANGE MANAGEMENT (1974), RANGELAND HYDROLOGY (1973), RANGELAND ENTOMOLOGY (1974), RANGELAND PLANT PHYSIOLOGY (1977), BENCHMARKS: A STATEMENT OF CONCEPTS AND POSITIONS (Rev. 1978), RANGELAND REFERENCE AREAS (1975), selected symposia, and abstracts of papers of the annual meetings held each February.

SOCIETY OF AMERICAN FORESTERS (SAF)
Address: 5400 Grosvenor Lane, Washington, D.C. 20014

SAF was founded in 1900 to advance the science, technology, education, and practice of professional forestry in America. Its membership includes foresters and scientists working in related fields. SAF publishes the Journal of Forestry, Forest Science, and SAF Proceedings (annually). It also published the TERMINOLOGY OF FOREST SCIENCE (1971).

THE SOIL CONSERVATION SOCIETY OF AMERICA (SCSA)
Address: 7515 Northeast Ankeny Road, Ankeny, Iowa 50021

SCSA was founded in 1943 to promote the development and advancement of the science and art of proper land management and conservation of soil, water, air, and related renewable natural resources. SCSA membership includes professional soil and water conservationists and others in fields related to conservation, management, and use of natural resources. Publications include the Journal of Soil and Water Conservation, RESOURCE CONSERVA-

TION GLOSSARY (1976), and technical monographs and educational booklets.

SOIL SCIENCE SOCIETY OF AMERICA (SSSA)
Address: 677 S. Segoe Road, Madison, Wisc. 53711

SSSA was founded in 1936 by merger of the American Soil Survey Association and the Soils Section of American Society of Agronomy. This organization is affiliated with the American Society of Agronomy and the Crop Science Society of America; its membership includes soil scientists, conservationists, engineers, and others interested in fundamental and applied soil sciences. SSSA proceedings are published as the Soil Science Society of America Journal. This organization cooperates with ASA on other publications of joint interest.

WEED SCIENCE SOCIETY OF AMERICA (WSSA) (formerly Association of Regional Weed Conferences to 1960, Weed Society of America to 1967)
Address: 309 West Clark Street, Champaign, Illinois 61820

The WSSA was founded in 1953 to promote research, teaching, and extension education in weed control and to distribute information on all aspects of weed science. Its membership includes biological and chemical scientists and engineers involved in weed control, regulatory agency personnel, and research and sales personnel from chemical and equipment industries.

Its publications include Weed Science, HERBICIDE HANDBOOK (1979), and the proceedings of regional weed control conferences held throughout the U.S.

THE WILDLIFE SOCIETY (TWS) (Society of Wildlife Specialists to 1937)
Address: 7101 Wisconsin Avenue, N.W., Washington, D.C. 20014

The Wildlife Society was established in 1936 to encourage management of wildlife resources on a sound ecological basis and to foster distribution of information on wildlife management. Its membership includes wildlife biologists and others interested in wildlife management and natural resource conservation.

The publications of this society include the Journal of Wildlife Management, Wildlife Society Bulletin, WILDLIFE MANAGEMENT TECHNIQUES (1971), and various wildlife monographs.

Chapter 5

SELECTED LITERATURE OF RANGE SCIENCE

KEY RANGE SCIENCE REFERENCE LITERATURE[47]

Blood, Don A. 1976. Rocky Mountain Wildlife. Hancock House Publishers, Saanichton, Br. Col. 127 p.

> Reference book on Rocky Mountain wildlife; well illustrated; part I, biology and ecology of major families and groups of mammals and birds; part II, facts presented in tabular form on key species of mammals.

Branson, Farrel A., Gerald F. Gifford, and J. Robert Owen. 1973. Rangeland Hydrology. Soc. Range Mgt. Range Sci. Ser. 1. 84 p.

> Hydrologic principles as applied to range ecosystems; emphasis given to water yields and to erosion and sedimentation in arid and semiarid regions; prepared by members of a sciential committee of the Society for Range Management; 274 entries included in a composite literature cited section.

Crampton, E.W., and L.E. Harris. 1969. (2nd Ed.). Applied Animal Nutrition. W.H. Freeman and Co., San Francisco. 753 p.

> Subtitle, "The Use of Feedstuffs in the Formulation of Livestock Rations"; attempts to bring together the theory of animal nutrition and the practice of animal feeding; an intermediate level textbook with range livestock applications; appendices include feed composition and glossary of feed terminology; chapters provided with short sections of suggested readings.

Dasmann, Raymond F. 1964. Wildlife Biology. John Wiley and Sons, New York. 231 p.

> Ecological basis of wildlife practices in terrestrial communities, primarily of range and forest; emphasis given to the principles of wildlife biology as a basis of applied wildlife management; a beginning college textbook or reference providing background information; includes 184 references.

Daubenmire, Rexford [F]. 1968. Plant Communities: A Textbook of Plant Synecology. Harper and Row, New York. 300 p.

An advanced college textbook emphasizing the study of plant communities as components of ecosystems; major sections as follows: the nature of plant communities, analysis and description of plant communities, plant succession, and vegetation and ecosystem classification; about 475 entries in a composite references section.

_____. 1974 (3rd Ed.). Plants and Environment--A Textbook of Plant Autecology. John Wiley and Sons, New York. 422 p.

A college textbook on autecology; emphasis given to a working knowledge of the basic interrelationships between the individual plant and its environment; individual chapters devoted to individual environmental factors with final chapters on the environmental complex and ecological plant adaption and evolution; provides background for understanding the autecology of range plants; 868 entries included in a single literature cited section.

Dayton, W.A. (Supv.). 1940 (Rev.). Range Plant Handbook. U.S. Govt. Printing Office, Washington, D.C. Variously paged.

Summarizes the identification and relative importance of key plants on western U.S. mountain ranges; includes 339 generic and specific write-ups with notes on 500 additional species arranged in sections on grass, grasslike, weed (forb), and browse plants; includes excellent drawings and descriptions of the key plants; prepared by U.S. Forest Service personnel; originally published in 1937, slightly revised in 1940; still useful but needs to be revised in areas of forage value and utilization methods; reprinted in paperback form by Clearinghouse, U.S. Dept. of Commerce, Washington, D.C.

Ensminger, M.E. 1970 (4th Ed.). Sheep and Wool Science. Interstate Print. and Pub., Danville, Ill. 948 p.

Standard textbook on sheep production with range implications; one chapter devoted exclusively to range sheep management; minimum documentation; selected references placed at end of chapters.

_____. 1976 (5th Ed.). Beef Cattle Science. Interstate Print. and Pub., Danville, Ill. 1,556 p.

Standard textbook on beef cattle production with range implications; chapters grouped into three parts: general beef cattle, cow-calf systems and stockers, and cattle feedlots and pasture fattening; composition of feeds table in appendix; minimum documentation; selected references placed at end of chapters.

Giles, Robert H., Jr. (Ed.). 1971 (3rd Ed., Rev.). Wildlife Management Techniques. The Wildl. Soc., Washington, D.C. 633 p.

Standard, intermediate-advanced textbook and manual on management of game animals and birds; goal to improve the management of the wildlife resource through more rapid development and improved use of resources; includes two extensive chapters on habitat analysis and manipulation; remaining chapters emphasize animal biology; prepared by Wildlife Techniques Manual Comm. with chapters individually authored; first published in 1969 but previously published in 1960 and 1964 under different titles; composite literature cited section of about 1900 entries.

Gray, James R. 1968. Ranch Economics. Iowa State Univ. Press, Ames. 534 p.

Treats ranching as a business subject to economic laws; attempts to bring together in one volume the economic theory and research that deals with western livestock ranges and to present the economic consequences of ranch management decisions; written for ranchers, students, and others associated with the range livestock industry; 233 entries included in chapter literature cited sections.

Heady, Harold F. 1975. Rangeland Management. McGraw-Hill Book Co., New York. 460 p.

Standard, intermediate level college textbook for range science; emphasis indicated by chapter groupings as follows: grazing animals as ecological factors (the grazing processes), management of grazing animals (numbers, kinds, distribution, season-systems), and management of vegetation (range improvements); author indicates that it was "not intended to be the text for courses in range ecology, range resources, range plants, range inventory and planning, range economics, or range policy"; worldwide coverage; 1,179 entries included in individual chapter literature cited sections.

Heath, Maurice E., Darrel S. Metcalfe, and Robert F. Barnes (Eds.). 1973 (3rd Ed.). Forages: The Science of Grassland Agriculture. Iowa State Univ. Press, Ames. 755 p.

Chapters written by 96 contributing editors and authors; emphasis given to cultivated forages of the United States including Alaska and Hawaii, but materials integrate with range seeding and utilization; 65 chapters arranged in four major divisions: forages and a productive agriculture, forage grasses and legumes, forage production practices, and forage utilization; first edition published in 1951 and second edition in 1962; about 2,600 citations included as footnotes.

Hewitt, George B., Ellis W. Huddleston, Robert J. Lavigne, Darrel N. Ueckert, and J. Gordon Watts. 1974. Rangeland Entomology. Soc. Range Mgt. Range Sci. Ser. 2. 127 p.

A single source of information about the entomological aspects of range management with western U.S. emphasis; presents summary data on the importance, problems, and control of insects in range ecosystems; intended as a supplementary college text for range management students and as an abbreviated reference manual on rangeland insects; excludes livestock ectoparasites; prepared by members of an SRM sciential committee; 416 selected references arranged by chapter.

Keeler, Richard F., Kent R. Van Kampen, and Lynn F. James (Eds.). 1978. Effects of Poisonous Plants on Livestock. Academic Press, New York. 600 p.

Emphasis given to gross and microscopic effects of poisonous plants on livestock, the identification of toxins, and the biochemical mechanisms of action; some papers emphasize the practical solutions of managing livestock to prevent or reduce poisoning; includes 56 papers presented by 77 U.S. and Australian authors at a U.S.-Australian Symposium on Poisonous Plants held at Logan, Utah, in 1977; all papers include moderate to extensive reference sections.

Kingsbury, John M. 1964. Poisonous Plants of the United States and Canada. Prentice-Hall, Englewood Cliffs, N.J. 626 p.

Presents information on U.S. and Canadian plants known to have poisoned livestock or humans; provides for each major plant or plant group the scientific name, common name or names, description, distribution and habitat, poisonous principle, toxicity symptoms, and conditions of poisoning; a reference book for veterinarians, students, and others working with poisonous plants; includes 1,715 references in a composite bibliography.

Kuchler, A.W. 1964. Potential Natural Vegetation of the Conterminous United States. Amer. Geogr. Soc. Spec. Pub. 36. 116 p.

A manual developed to accompany and explain a map with the same title (39" x 60", scale of 1:3,168,000 or 1" = about 50 mi.); presents a system for classifying the natural vegetation of the U.S. excluding Alaska and Hawaii; provides a photo and brief description (the physiognomic dominants, other components, and geographical occurrence) in outline form for each vegetation type; includes a selected bibliography of 318 entries on plant geography and synecology of the U.S.

McCall, Joseph R., and Virginia N. McCall. 1977. Outdoor Recreation--Forest, Park and Wilderness. Benziger Bruce and Glencoe, Inc., Beverly Hills, Calif. 358 p.

An introductory-intermediate textbook on outdoor recreation; emphasis given to park administration and management but with some extension into recreational use of extensive rangelands; chapters organized into sections on the beginnings of outdoor recreation, the recreation landlords, and managing the resources; minimal literature citations are found as footnotes.

McKell, Cyrus M., James P. Blaisdell, and Joe R. Goodin (Eds.). 1972. Wildland Shrubs--Their Biology and Utilization. USDA, For. Serv. Gen. Tech. Rep. INT-1. 494 p.

A review of the important aspects of shrub biology and utilization throughout the world; chapters grouped by sections: continental aspects, present and possible future uses, genetic potential, synecology, physiology, nutritive quality, and regeneration; a symposium of papers presented at an international conference at Logan, Utah, in 1971; a composite references section (p. 445-94) of about 1,750 items.

National Research Council, Subcommittee on Beef Cattle Nutrition. 1976 (5th Rev. Ed.). Nutrient Requirements of Domestic Animals. Number 4. Nutrient Requirements of Beef Cattle. Natl. Acad. Sci., Washington, D.C. 56 p.

Emphasis given to nutrient requirements of beef cattle, means of meeting these requirements, and symptoms of nutrient deficiency; appendix includes tables of nutrient requirements and feed composition; prepared by a subcommittee with Tilden Wayne Perry as chairman; includes a selected bibliography of 98 items.

National Research Council, Subcommittee on Range Research Methods. 1962. Basic Problems and Techniques in Range Research. Natl. Acad. Sci.-Natl. Res. Council Pub. 890. 341 p.

Discusses the problems inherent to range research; assembles the various range animal, plant, and land research methods and describes their use, limitations, and suitability; a reference guide to range research methodology and textbook in range research, but now out of print; many procedures have management uses; prepared by six committee members (C. Wayne Cook, Chm.) with 61 contributors; extensive literature sections accompany most of the 11 chapters.

National Research Council, Subcommittee on Sheep Nutrition. 1975 (5th Rev. Ed.). Nutrient Requirements of Domestic Animals. Number 5. Nutrient Requirements of Sheep. Natl. Acad. Sci., Washington, D.C. 72 p.

Emphasis given to nutrient requirements of farm and range sheep, means of meeting these requirements, and symptoms of nutrient deficiency; includes tables of nutrient requirements and feed composition; prepared by a subcommittee with Arthur L. Pope as chairman; first edition published in 1945; bibliographic section includes 339 items.

O'Mary, Clayton C., and Irwin A. Dyer (Eds.). 1978. Commercial Beef Cattle Production. Lea and Febiger, Philadelphia, Penn. 414 p.

Emphasis given to applied aspects of beef cattle production including range aspects; first chapters deal with planning beef cattle operations; the following chapters cover various aspects of management and the interrelationships within the beef cattle industry; final two chapters deal with beef cattle organizations and the use of educational media in classroom presentations; all chapters written by contributing authors and include a list of references.

Paulsen, Harold A., Jr., and Elbert H. Reid (Eds.). 1970. Range and Wildlife Habitat Evaluation--A Research Symposium. USDA Misc. Pub. 1147. 220 p.

Over half of publication deals with forage evaluation for livestock and wildlife, the remainder with remote sensing in range and wildlife habitat evaluation and range and wildlife habitat ecology; includes 41 contributors and based on a conference held in Arizona in 1968 and sponsored by the U.S. Forest Service; includes chapter literature citations.

Public Land Law Review Commission. 1970. One Third of the Nation's Land. U.S. Govt. Printing Office, Washington, D.C. 342 p.

The report of a presidential commission charged with formulating recommendations for managing the federal lands in the U.S., comprising about 1/3 of its land area; intended to represent a broad consensus on basic land management principles and recommendations for carrying them out; special attention given to and background information provided on range and related resources of timber, minerals, water, fish and wildlife, outdoor recreation, and intensive agriculture.

Sampson, Arthur W. 1951. Range Management, Principles and Practices. John Wiley and Sons, New York. 570 p.

Intended as a comprehensive text for the broad field of range management but now out of print; organized into sections on range management in perspective, native range forage plants, improvement and management of range and stock, and protection of range resources and range livestock; represents the range management philosophy of a long-time authority in the field of range and management; preceded by earlier books by the same author on RANGE AND PASTURE MANAGEMENT (1923), NATIVE AMERICAN PLANTS (1924), and LIVESTOCK HUSBANDRY ON RANGE AND PASTURE (1928); literature cited sections by chapter.

Scott, George E. 1975 (2nd Ed.). The Sheepman's Production Handbook. Sheep Industry Development Program, 200 Clayton St., Denver, Colo. 246 p.

A practical production manual divided into sections on genetics, reproduction, health, nutrition, management, and marketing; a reference book and textbook for students, producers, researchers, and allied industry personnel; first published in 1970.

Sosebee, Ronald E. (Ed.). 1977. Rangeland Plant Physiology. Soc. for Range Mgt. Range Sci. Ser. 4. 290 p.

A review of selected topics of plant physiology as a basis for managing range plants and rangeland resources generally; deals with "plant physiology, or perhaps more appropriately plant ecophysiology, the physiological response of the vegetation resources to the environment to which they are adapted"; prepared by a special SRM sciential committee; 1239 literature citations arranged in chapter sections.

Stoddart, Laurence A., Arthur D. Smith, and Thadis W. Box. 1975 (3rd Ed.). Range Management. McGraw-Hill Book Co., New York. 532 p.

Standard, intermediate level textbook in range management covering most significant topics in the field except range plants and range economics; first edition published in 1943, second edition in 1955; third edition revised by Smith and Box and gives greater attention to worldwide range science; 1,099 bibliographic entries arranged in chapter sections.

Thames, John L. (Ed.). 1977. Reclamation and Use of Disturbed Land in the Southwest. Univ. Ariz. Press, Tucson. 362 p.

Deals with the sociological and biological impact of mining on other land values and the reclamation of mining-disturbed lands; the first three sections are of a general nature; the last two sections deal specifically with rehabilitation and revegetation techniques for disturbed lands; authored by 41 contributors and evolved from a symposium held at Tucson, Arizona, in 1975.

U.S. Dept. Agric. 1948. Grass, The Yearbook of Agriculture, 1948. U.S. Govt. Printing Office, Washington, D.C. 892 p.

Represents the contributions of many scientists in USDA and the land grant colleges to the subject of grass; now considerably outdated but a land mark in range and related literature and philosophy; first section covers the broad uses of grasses in conservation, pasture, range, turf, and harvested forages; second section deals with regional considerations on grass production and utilization; third section emphasizes the identification and evaluation of individual grasses, legumes, and associated plants.

USDA, Forest-Range Task Force. 1972. The Nation's Range Resources--A Forest-Range Environmental Study. USDA, For. Serv. Resource Rep. 19. 147 p.

An inventory and evaluation of all range and forest lands in the 48 conterminous states with actual or potential value for livestock grazing; "forest-range" categorized into 34 major ecosystems (based on Kuchler) within four ecogroups and into three ownerships; projects future grazing needs and potential production under six levels of management; provides quantitative and qualitative data on "forest-range" by ecosystem, ecogroup, state, land ownership, and present levels of management; projects a model for range resources allocation and decision making.

Vallentine, John F. 1971. Range Development and Improvements. Brigham Young Univ. Press, Provo, Utah. 516 p.

Principles and practices of plant control, revegetation, mechanical treatments, fertilization, rodent and insect control, and range animal handling facilities on rangelands; an intermediate-advanced textbook and reference manual emphasizing regional application in North America; written for use by students, educators, and practitioners; special appendices on partial budgeting, properties of herbicides, chemical control recommendations, and forage plants for range seeding; 1,202 literature citations arranged by chapter sections.

Van Dyne, George M. (Ed.). 1969. The Ecosystem Concept in Natural Resources Management. Academic Press, New York. 383 p.

Emphasizes natural resources ecosystems as related to range, forest, watershed, fishery, and wildlife resource management; a useful reference in college level resource management courses; includes the contributions of 13 authors and based on a symposium held in 1968 and sponsored by the Amer. Soc. Range Mgt.; references provided at the end of each chapter.

Younger, V.B., and C.M. McKell (Eds.). 1972. The Biology and Utilization of Grasses. Academic Press, New York. 426 p.

Objective was to review the knowledge about grass biology and to provide a broader understanding of the important role of grasses in man's existence; contents include five chapters on grass evolution and genetics, six chapters on vegetative growth and plant development, five chapters on the ecology of grasses, four chapters on soils and mineral nutrition, and individual chapters on effects of defoliation, carbohydrate reserves, physiology of flowering, and seed production; contributions made by 30 authors; based on a symposium held at Riverside, Calif., in 1969.

SUBJECT BIBLIOGRAPHY OF NORTH AMERICAN RANGE SCIENCE [48]

Range Plants

Beath, O.A., C.S. Gilbert, H.F. Eppson, and Irene Rosenfeld. 1953. Poisonous Plants and Livestock Poisoning. Wyo. Agric. Expt. Sta. Bul. 324. 94 p.

Covers subjects of poisonous plants, toxic minerals, and methods of controlling and preventing livestock poisoning, with special reference to Wyoming.

Beetle, Alan A. 1970. Recommended Plant Names. Wyo. Agric. Expt. Sta. Res. J. 31. 124 p.

Recommends scientific names and common names for use with range plants of the United States; arranged alphabetically by scientific name; includes a common name index to the scientific name genera.

Beetle, Alan A., and Morton May. 1971. Grasses of Wyoming. Wyo. Agric. Expt. Sta. Res. J. 39. 151 p.

Description, habitat, and distribution of native and selected introduced grasses in Wyoming; distribution maps for native species; keys to tribes, to genera, and sometimes to species; scientific and common name indexes.

Blauer, A. Clyde, A. Perry Plummer, E. Durrant McArthur, Richard Stevens, and Bruce C. Giunta. 1975. Characteristics and Hydridization of Important Intermountain Shrubs. I. Rose Family. USDA, For. Serv. Res. Paper INT-169. 36 p.

Description, hybridization, distribution, habitat, and use of important Intermountain rosaceous shrubs; includes pioneer hybridization studies; key to genera and species.

______. 1976. Characteristics and Hybridization of Important Intermountain Shrubs. II. Chenopod Family. USDA, For. Serv. Res. Paper INT-177. 42 p.

Description, hybridization, distribution, habitat, and use of important Intermountain chenopod shrubs; includes results of interspecific and intergeneric hybridization experiments; key to genera and species.

Campbell, J.B., K.F. Best, and A.C. Budd. 1956. Ninety-nine Forage Plants of the Canadian Prairies. Can. Dept. Agric. Pub. 964. 99 p.

Describes habitat, growth characters, nutritive values, palatability, reaction to grazing, and drought tolerance plus line drawings of the principal forage plants of the Canadian prairies.

Campbell, Robert S., and Carlton H. Herbel (Eds.). Improved Range Plants. Soc. Range Mgt. Range Symposium Ser. 1. 90 p.

Papers dealing with the present status and future opportunities for improving range plants through selection and breeding; proceedings of a symposium held Feb. 5, 1974, in conjunction with the annual meeting of the Society for Range Management.

Cook, C. Wayne. 1966. The Role of Carbohydrate Reserves in Managing Range Plants. Utah Agric. Expt. Sta. Mimeo. Ser. 499. 11 p.

A summarization of the importance of plant food reserves and their implications in grazing management; illustrations relate herbage yield and carbohydrate levels.

Copple, R.F., and C.P. Pase. 1978. A Vegetative Key to Some Common Arizona Range Grasses. USDA, For. Serv. Gen. Tech. Rep. RM-53. 106 p.

Detailed description of vegetative characters plus sketch of ligule-collar area of 77 common Arizona range grasses; includes vegetative keys, also explanation and glossary of vegetative characters.

Crampton, Beecher. 1961 (Rev. Ed.). Range Plants--A Laboratory Manual. Univ. Calif., Dept. Agron., Davis. 210 p.

Key to the more common range plants of California and adjacent areas with information inserted on description, distribution and habitat, and value for many species; not illustrated.

_____. 1974. Grasses in California. Univ. Calif. Press, Berkeley. 178 p.

Key to the grasses with information inserted on description, distribution and habitat, and value for many species; illustrated by line drawings and color pictures.

Crawford, Hewlette S., Clair L. Kucera, and John H. Ehrenreich. 1969. Ozark Range and Wildlife Plants. USDA Agric. Handbook 356. 236 p.

Distribution and site, description (including line drawing), and importance for livestock and wildlife of common range plants of the Ozark region; includes general description of the region, summer and winter keys to the plant genera, glossary, and selected literature section.

Dayton, William A. 1938. Important Western Browse Plants. USDA Misc. Pub. 101. 213 p.

Description, distribution, and importance of native browse plants in western U.S.; arranged by family, genus, and species; well illustrated by drawings and pictures; out of print; particularly useful in species identification and distribution; comprehensive plant name index. Originally published in 1931.

_____. 1960. Notes on Western Range Forbs: Equisetaceae Through Fumariaceae. USDA Agric. Handbook 161. 254 p.

Emphasis given to important and abundant range forbs in the 11 western states, Equisetaceae through Fumariaceae only; materials arranged by family and then genus and species; comprehensive plant name index; few species illustrated.

Durrell, L.W., Rue Jensen, and Bruno Klinger. 1952 (Rev.). Poisonous and Injurious Plants in Colorado. Colo. Agric. Expt. Sta. Bul. 412. 88 p.

Descriptions (with drawings), conditions of poisoning, livestock symptoms, and remedial actions suggested in relation to the principal poisonous and injurious plants of Colorado; plant index and poisonous property index.

Evers, Robert A., and Roger P. Link. 1972. Poisonous Plants of the Midwest and Their Effects on Livestock. Univ. of Ill. (Urbana, Ill.) Spec. Pub. 24. 165 p.

Based on the poisonous plants of Illinois and their description, poisoning effects, and prevention of poisoning; well illustrated.

Freeman, John D., and Harold D. Moore. 1974. Livestock-Poisoning Vascular Plants of Alabama. Ala. Agric. Expt. Sta. Bul. 460. 79 p.

Directed toward the recognition of poisonous plants and the reduction of livestock losses; includes general and plant species sections; plant species illustrated in color; appendix tables, glossary, and index.

Furniss, Malcolm M., and William F. Barr. 1975. Insects Affecting Important Native Shrubs of the Northwestern United States. USDA, For. Serv. Gen. Tech. Rep. INT-19. 64 p.

Emphasis given to geographic range, hosts, type of damage, appearance and habits, life cycle, and natural control of 43 insect species or groups affecting browse plants of the Pacific Northwest.

Gay, Charles W., Jr., and Don D. Dwyer. 1967. Poisonous Range Plants. N. Mex. Agric. Ext. Cir. 391. 21 p.

Description, habitat, conditions of poisoning, and prevention and treatment in relation to the common poisonous plants and associated livestock poisoning in New Mexico.

______. 1970. New Mexico Range Plants. N. Mex. Agric. Ext. Cir. 374. 86 p.

Description (with drawings), occurrence (with distribution map), and forage values and management of selected, important New Mexico range plants; includes an introduction to rangeland areas of New Mexico; plant name indexes.

Gill, John D., and William M. Healy. 1974. Shrubs and Vines for Northeastern Wildlife. USDA, For. Serv. Gen. Tech. Rep. NE-9. 180 p.

Range, habitat, life history, uses, propagation, and management but not identification of 100 shrubs and woody vines useful for wildlife in the Northeast; good literature cited section; not illustrated.

Gould, Frank W. 1968. Grass Systematics. McGraw-Hill Book Co., New York. 382 p.

Designed as a reference manual in agrostology; emphasis given to new evidence on grass phylogeny; includes sections on the anatomy of the grass plant, reproduction and cytogenetics, and revised grass classification system, grass genera of the United States, and grassland associations in North America; appendixes on herbarium procedures and grass nomenclature; glossary; index.

Grelen, Harold E., and Vinson L. Duvall. 1966. Common Plants of Longleaf Pine-Bluestem Range. USDA, For. Serv. Res. Paper SO-23. 96 p.

Description, food value for cattle and wildlife, and geographic range of the common range plants of this vegetation type from East Texas to Florida; well illustrated; emphasizes vegetative plant identification; glossary; plant index.

Halls, L.K., and T.H. Ripley (Eds.). 1961. Deer Browse Plants of Southern Forests. USDA, For. Serv., Southern and Southeast For. Expt. Sta. 78 p.

Identification, distribution and habitat, and value of the principal deer browse plants of Southern forests; illustrated.

Hamilton, Harry (Ed.). 1964. Forage Plant Physiology and Soil-Range Relationships. Amer. Soc. Agron. (Madison, Wisc.) Pub. 5. 250 p.

Papers presented at a symposium held at Denver in 1963; emphasis given to the physiology, autecology, and productivity of pasture and range.

Hardin, James W. 1973 (Rev.). Stock-Poisoning Plants of North Carolina. N. Car. Agric. Expt. Sta. Bul. 414. 138 p.

Comprehensive coverage of livestock poisoning plants of North Carolina, including their description, habitat and distribution (with illustration), conditions of poisoning, and prevention and treatment; plants partially illustrated; glossary; plant index.

Haws, B. Austin. 1978. Economic Impacts of Labops hesperius on the Production of High Quality Range Grasses. Utah Agric. Expt. Sta., Logan, Utah. 269 p.

Studies on the biological effects and economic impacts of black grass bugs on native and seeded range grasses.

Hayes, Doris W., and George A. Garrison. 1960. Key to Important Woody Plants of Eastern Oregon and Washington. USDA Agric. Handbook 148. 227 p.

Includes spring-summer and fall-winter keys for deciduous plants and a key to persistent-leaved plants, frequently with additional species description and distribution; well illustrated; notes on forage value; plant name index.

Hermann, F[rederick].J. 1966. Notes on Western Range Forbs: Cruciferae Through Compositae. USDA Agric. Handbook 293. 365 p.

Important and abundant range forbs in the 11 western states, Cru-

ciferae through *Compositae* only; continuation of USDA Agric. Handbook 161; arranged by families and then genus and species; illustrated; comprehensive plant name index.

______. 1970. Manual of the Carices of the Rocky Mountains and Colorado Basin. USDA Agric. Handbook 374. 397 p.

Detailed description (including line drawing) of 164 selected species plus distribution for each and sometimes forage value and uses; key to *Carex* sections and key to *Carex* species; glossary; species index.

Herzman, Carl W., A.C. Everson, Myron H. Mickey, Ivan R. Porter, et al. 1958. Handbook of Colorado Native Grasses. Colo. Agric. Ext. Bul. 450. 31 p.

Popular treatise of 44 Colorado native grasses, giving drawing, brief description, habitat, and uses and values for each.

Hitchcock, A.S., and Agnes Chase. 1951 (Rev.). Manual of the Grasses of the United States. USDA Misc. Pub. 200. 1051 p.

Comprehensive keys to tribes and species; descriptions and distributions for native grasses and a few introduced grasses of the United States; extensive synonymy section; comprehensive scientific name-common name index; well illustrated by line drawings and distribution maps; reprinted in two volumes by Dover Publications, New York, 1971.

Hoffman, G.O., and B.J. Ragsdale. 1966. Know Your Grasses. Texas Agric. Ext. Bul. 182. 47 p.

Popular treatise of important Texas grasses; includes description, line drawings, distribution, and evaluation of each; distribution tied in with vegetational areas.

Humphrey, Robert R. 1958. Arizona Range Grasses. Ariz. Agric. Expt. Sta. Bul. 298. 104 p.

Description (including line drawings), forage value, and management of important Arizona range grasses; arranged by scientific names; common name index.

Johnson, James R., and James T. Nichols. 1970. Plants of South Dakota Grasslands. S. Dak. Agric. Expt. Sta. Bul. 566. 163 p.

Covers the primary species of South Dakota plains, prairies, and associated tame pastures; includes description (including color photograph), distribution, and uses and values for each; glossary; common name-scientific name index.

Judd, B. Ira. 1962. Principal Forage Plants of Southwestern Ranges. Rocky Mtn. For. and Range Expt. Sta. Paper 69. 93 p.

Description (including line drawings), distribution from western Texas through Arizona, and uses and values; good coverage of rangeland types in the Southwest.

Kreitlow, Kermit W., and Richard H. Hart (Coord.). 1974. Plant Morphogenesis as the Basis of Scientific Management of Range Resources: Proceedings of the Workshop of the United States-Australia Rangelands Panel, Berkeley, Calif., March 29-April 5, 1971. USDA Misc. Pub. 1271. 232 p.

Physiology, morphology, and management of range plants; grazing effects on range plants and range plant communities.

Lamb, Samuel H. 1971. Woody Plants of New Mexico and Their Value to Wildlife. N. Mex. Dept. Game and Fish Bul. 14. 80 p.

Identification (mostly limited to pictures and drawings), ecology, and wildlife value of trees and shrubs in New Mexico; arranged by plant species groups and by vegetational zones; well illustrated; indexes by plant names and uses.

Leininger, Wayne C., John E. Taylor, and Carl L. Wambolt. 1977. Poisonous Range Plants in Montana. Mon. Agric. Ext. Bul. 348. 60 p.

A practical guide for reducing livestock losses from poisonous plants; part 1 covers general principles; part 2 covers the principal poisonous plant species in Montana, including their description, ecology, conditions of poisoning, and symptoms; color illustrations; an appendix table includes all known or suspected livestock-poisoning plants in Montana.

Leithead, Horace L., Lewis L. Yarlett, and Thomas N. Shiflet. 1971. 100 Native Forage Grasses in 11 Southern States. USDA Agric. Handbook 389. 216 p.

Grasses selected on basis of forage value and range indicator value; provides description, growth characteristics, distribution, site adaptation, and use and management for each; well illustrated by line drawings and distributional maps; glossary; plant name index.

Lodge, Robert W., Alastair McLean, and Alexander Johnston. 1968. Stock-poisoning Plants of Western Canada. Can. Dept. of Agric. Pub. 1361. 34 p.

Recognition of poisonous plants, plant control, animal management to avoid poisoning, and animal treatment; emphasis given to major poisonous species.

Morris, H.E., W.E. Booth, G.F. Payne, and R.E. Stitt. 1954. Important Grasses on Montana Ranges. Mon. Agric. Expt. Sta. Bul. 500 58 p.

Description (with illustrations), use, and distribution of the principal grass species; common name index and scientific name index.

Nickerson, Mona F., Glen E. Brink, and Charles Feddema. 1976. Principal Range Plants of the Central and Southern Rocky Mountains: Names and Symbols. USDA, For. Serv. Gen. Tech. Rep. RM-20. 121 p.

Checklist of range plants; arranged by scientific names within grass, grasslike, forb, and tree-shrub sections; accepted and other common names provided; index of common names.

Norris, J.J., and K.A. Valentine. 1954. Principal Livestock-poisoning Plants of New Mexico Ranges. N. Mex. Agric. Expt. Sta. Bul. 390. 78 p.

An older but rather comprehensive coverage of poisonous plants; arranged alphabetically by common name; includes a key to plants based on symptoms, habitats, and distribution; common name index.

Oefinger, Simeon W., Jr., and Lowell K. Halls. 1974. Identifying Woody Plants Valuable to Wildlife in Southern Forests. USDA, For. Serv. Res. Paper SO-92. 76 p.

Semitechnical description of 70 selected woody plants; twigs, buds, and other key identification features illustrated in color; index to scientific names.

Ohlenbusch, Paul D. 1976. Range Grasses of Kansas. Kan. Agric. Ext. Cir. 567. 20 p.

Popular, abridged, illustrated description and evaluation of selected range grasses of Kansas.

Phillips Petroleum Co. 1963. Pasture and Range Plants. Bartlesville, Okla. 176 p.

Description (including color pictures) and uses and values of selected native and introduced grasses and forbs and poisonous plants; pertains primarily to the prairie region of central U.S.; a composite of six previous booklets.

Schmutz, Ervin M., Barry N. Freeman, and Raymond E. Reed. 1968. Livestock-poisoning Plants of Arizona. Univ. Ariz. Press, Tucson. 176 p.

Description, distribution and habitat, conditions of poisoning, and prevention of poisoning associated with the principal livestock poisoning plants of Arizona; illustrated.

Smith, C. Earle, Jr. 1971. Preparing Herbarium Specimens of Vascular Plants. USDA Agric. Info. Bul. 348. 29 p.

Collecting, recording, pressing, drying and storing plant specimens; special section on hard-to-press specimens.

Sperry, O.E., J.W. Dollahite, G.O. Hoffman, and B.J. Camp. 1964. Texas Plants Poisonous to Livestock. Texas Agric. Expt. Sta. Bul. 1028. 59 p.

Livestock poisoning problems and the description (with pictures), distribution, conditions of poisoning, symptoms caused, and management and treatment related to the major poisonous plants of Texas; references; plant index.

Stoddart, L.A., A.H. Holmgren, and C. Wayne Cook. 1949. Important Poisonous Plants of Utah. Utah Agric. Expt. Sta. Spec. Rep. 2. 21 p.

Recognition and prevention of losses from the dangerous poisonous plants of the state; color pictures and line drawings.

Stutz, Howard C. (Ed.). 1975. Wildland Shrubs Symposium and Workshop Proceedings. USDA, For. Serv., Shrub Sciences Lab., Provo, Utah. 168 p.

Papers on current wildland shrub research presented in conjunction with the dedication of the U.S. Forest Service Shrub Sciences Laboratory, Provo, Utah, in 1976.

USDA, Agric. Res. Serv. 1968. 22 Plants Poisonous to Livestock in the Western States. USDA Agric. Info. Bul. 327. 64 p.

Recognition, effects, and reducing losses from the major poisonous plants of the West; color illustrations and distribution maps.

USDA, For. Serv. 1975. Some Important Native Shrubs of the West. USDA, For. Serv. Intermtn. For. and Range Expt. Sta. 16 p.

Popular introduction with color pictures to 17 selected native shrubs of the Intermountain West.

Vallentine, John F. 1961. Important Utah Range Grasses. Utah Agric. Ext. Cir. 281. 48 p.

Identification (with line drawings), distribution, and forage value of 55 native and introduced range grasses; description and color map of range types of Utah; common name index; emphasizes forage production and soil stabilization.

______. 1967. Nebraska Range and Pasture Grasses. Neb. Agric. Ext. Cir. 67-170. 55 p.

Identification (with line drawings), distribution, and uses and values of 59 range and pasture grasses of Nebraska; description of principal range sites in Nebraska; plant name index.

Wagner, Warren L., and Earl F. Aldon. 1978. Manual of the Saltbushes (Atriplex spp.) in New Mexico. USDA, For. Serv. Gen. Tech. Rep. RM-57. 50 p.

A manual for the identification and distribution of saltbushes commonly found in New Mexico.

Yarlett, Lewis L. 1965. Important Native Grasses for Range Conservation in Florida. USDA, Soil Cons. Serv., Gainesville, Fla. 163 p.

Identification (including pictures), distribution and site adaptation, growth habits, and forage value and management of the principal native grasses; grazing practices; checklist of all Florida grasses.

Young, James A., Raymond A. Evans, Burgess L. Kay, Richard E. Owen, and Frank L. Jurak. 1978. Collecting, Processing and Germinating Western Wildland Plants. USDA Agric. Reviews and Manuals. ARM-W-3. 38 p.

Techniques developed from practical experience and adapted from related fields as applied to western wildland plants; literature citations; glossary.

Range Resources and Ecology

Aldon, Earl F., and Thomas J. Loring (Tech. Coord.). 1977. Ecology, Uses, and Management of Pinyon-Juniper Woodlands: Proceedings of the Workshop. USDA, For. Serv. Gen. Tech. Rep. RM-39. 48 p.

Status of knowledge of the pinyon-juniper vegetation type of the Southwest; 12 papers; proceedings of a workshop held at Albuquerque, N. Mex., in 1977.

Austin, Morris E. 1972. Land Resource Regions and Major Land Resource Areas of the United States. USDA Agric. Handbook 296. 82 p.

Color map with supporting descriptions of resource areas in U.S. exclusive of Alaska and Hawaii; emphasis on soil resources; developed primarily for use by the Soil Conservation Service in developing soil and water conservation programs.

Bird, Ralph D. 1961. Ecology of the Aspen Parkland of Western Canada. Can. Dept. Agric. Pub. 1066. 155 p.

Vegetation, environmental factors affecting, and conservation of aspen parkland sites; well illustrated.

Blaisdell, James P., Vinson L. Duvall, Robert W. Harris, R. Duane Lloyd, Richard J. McConnen, and Elbert H. Reid. 1970. Range Ecosystem Research--The Challenge of Change. USDA Agric. Info. Bul. 346. 26 p.

The present and projected future status of rangeland in the U.S. and implications for future research on the problems of managing, conserving, and using rangelands.

Bose, Dan R. 1978. Rangeland Resources of Nebraska. USDA, Soil Cons. Serv., Lincoln, Neb. 121 p.

An inventory and description of Nebraska's rangeland resources, their present status and productivity, and their present and future uses and values.

Brockman, C. Frank, and Lawrence C. Merriam, Jr. 1973 (2nd Ed.). Recreational Use of Wildlands. McGraw-Hill Book Company, New York. 329 p.

Practical approaches to outdoor recreation on wildlands including rangelands.

Campbell, Robert S., and Wesley Keller. 1973. Range Resources of the Southeastern United States. Amer. Soc. Agron. (Madison, Wisc.) Spec. Pub. 21. 78 p.

Description, ecology, productivity, and integrated management of range resources in the Southeast; papers presented at a symposium at Miami Beach in 1972.

Can. Dept. Agric., Canada Soil Survey Comm. 1978. The Canadian System of Soil Classification. Can. Dept. Agric. Pub. 1646. 164 p.

Instructions for standardizing soil survey procedures in Canada; illustrated.

Clawson, Marion, and Burnell Held. 1957. The Federal Lands: Their Use and Management. The Johns Hopkins Press, Baltimore, Md. 539 p.

The origin, status, and uses of the federal lands; the policies, pricing processes, and revenues and expenditures associated with these lands; a 1957 summation.

Clayton, J.S., W.A. Ehrlich, D.B. Cann, J.H. Day, and I.B. Marshall. 1977. Soils of Canada; Vol. 1, Soil Report; Vol. 2, Inventory. Supply and Services Canada, Ottawa. 243 p., 239 p.

Dasmann, Raymond F. 1976 (Fourth Ed.). Environmental Conservation. John Wiley and Sons, New York. 436 p.

An introductory textbook on the ecology, conservation, and management of wildlands and natural resources.

Daubenmire, R[exford].F. 1970. Steppe Vegetation of Washington. Wash. Agric. Expt. Sta. Tech. Bul. 62. 131 p.

Vetetation components and climatic and soil data of 40 habitat types in the steppe region of eastern Washington; includes key to major steppe habitat types; illustrated.

_____. 1978. Plant Geography with Special Reference to North America. Academic Press, New York. 1978. 352 p.

Part 1, floristic plant geography--phylogenetic and historical; part 2, ecologic plant geography--sociologic and physiologic.

Daubenmire, R[exford].[F.], and Jean B. Daubenmire. 1968. Forest Vegetation of Eastern Washington and Northern Idaho. Wash. Agric. Expt. Sta. Tech. Bul. 60. 104 p.

Vegetation components and climatic and soil data of major forest habitat types; includes key to major forest habitat types; illustrated.

Davis, Kenneth P. 1976. Land Use. McGraw-Hill Book Co., New York. 324 p.

Textbook on land use principles, problems, and planning.

Ellison, Lincoln, A.R. Croft, and Reed W. Bailey. 1951. Indicators of Condition and Trend on High Range-watersheds of the Intermountain Region. USDA Agric. Handbook 19. 66 p.

Ecological principles, management implications, and indicators for judging range condition and trend.

Foss, Phillip O. 1960. Politics and Grass: The Administration of Grazing on the Public Domain. Univ. Wash. Press, Seattle, Wash. 235 p.

The origin of the public domain (lands administered by the Bureau of Land Management), the history and development of grazing practices and policies, and the administration of these lands.

Franklin, Jerry F., and C.T. Dyrness. 1973. Natural Vegetation of Oregon and Washington. USDA, For. Serv. Gen. Tech. Rep. PNW-8. 417 p.

The major natural vegetational units of Oregon and Washington and a description of the plant composition, environmental features, and successional patterns; illustrated; indexed; references.

Garrison, George A., Ardell J. Bjugstad, Don A. Duncan, Mont E. Lewis, and Dixie R. Smith. 1977. Vegetation and Environmental Features of Forest and Range Ecosystems. USDA Handbook 475. 68 p.

A system developed by the Forest-Range Environmental Study (FRES) for classifying forest-range of the 48 contiguous states into 34 ecosystems; each ecosystem described and illustrated; color map of forest and range ecosystems of the U.S.; good references section in appendix.

Heller, Robert C. 1975. Evaluation of ERTS-1 Data for Forest and Rangeland Surveys. USDA, For. Serv. Res. Paper PSW-112. 67 p.

Results of tests of using ERTS photographs in forest and range inventory.

Humphrey, Robert R. 1958. The Desert Grassland--A History of Vegetational Change and an Analysis of Causes. Ariz. Agric. Expt. Sta. Bul. 299. 62 p.

Historical changes in the desert grasslands of southwestern Texas to southern Arizona and their apparent causes.

______. 1962. Range Ecology. The Ronald Press Co., New York. 234 p.

A brief textbook on the ecology of natural grazing lands of the U.S.; chapters arranged by environmental factors with a special chapter on range condition; bibliography.

Humphrey, Robert R. (II-V), and Robert A. Darrow (I). 1944-1960. Arizona Range Resources (title varies): I. Cochise County; II. Yavapai County; III. Mohave County; IV. Coconino, Navajo, Apache Counties; V. Pima, Pinal, and Santa Cruz Counties. Ariz. Agric. Expt. Sta. Tech. Bul. 103, Bul. 299, 244, 266, 302. 56 p., 55 p., 79 p., 84 p., 138 p.

Regional approaches to the vegetation types, plant descriptions and forage values, range condition and trend, and grazing management in Arizona; range type maps; well illustrated.

Klingebiel, A.A., and P.H. Montgomery. 1973 (Rev.). Land-capability Classification. USDA Agric. Handbook 210. 21 p.

Criteria for placing agricultural lands into eight capability classes based on suitability to produce cultivated crops without deterioration; useful for classifying ranch lands when rangelands are additionally classified by range sites.

Kuchler, A.W. 1967. Vegetation Mapping. Ronald Press, New York. 472 p.

Basic considerations, technical aspects, and methodology of vegetation mapping and the application of vegetation maps.

Laycock, W.A. 1975. Rangeland Reference Areas. Soc. Range Mgt. Range Sci. Ser. 3. 66 p.

Need for preserving rangeland references areas; current programs of preserving reference areas; guideline for establishing reference areas; use of ecological baseline data in range management; literature review.

Lewis, Clifford E., Harold E. Grelen, Larry D. White, and Clifford W. Carter. 1974 (Rev.). Range Resources of the South. Ga. Agric. Expt. Sta. Bul. N.S. 9. 33 p.

Vegetation and uses of the major range types of the South; range type map; lists major centers of range information.

Lund, H. Gyde, Vernon J. LaBau, Peter F. Ffolliott, and David W. Robinson (Tech. Coord.). 1978. Integrated Inventories of Renewable Natural Resources: Proceedings of the Workshop. USDA, For. Serv. Gen. Tech. Rep. RM-55. 482 p.

Information requirements, current collection techniques, data handling, and progress being made in integrating inventories of renewable natural resources including range resources; papers presented at a workshop held January 8-12, 1978, at Tucson.

McLean, Alastair, and Leonard Marchand. 1968. Grassland Ranges in the Southern Interior of British Columbia. Can. Dept. Agric. Pub. 1319. 28 p.

Description, evaluation, and management of the principal range sites; general principles of range management; color illustrations.

Martin, Robert E., J. Edward Dealy, and David L. Caraher. 1978. Proceedings of the Western Juniper Ecology and Management Workshop, Bend, Oregon, January 1977. USDA, For. Serv. Gen. Tech. Rep. PNW-74. 177 p.

Papers from a symposium on western juniper (Juniperus occidentalis) emphasizing invasion, effect on range productivity, benefits, and control.

Natl. Res. Council, Comm. on Remote Sensing for Agric. Purposes. 1970. Remote Sensing With Special Reference to Agriculture and Forestry. Natl. Acad. Sci., Washington, D.C. 424 p.

Concepts, procedures, potentials and limitations, and applications of remote sensing.

Nielsen, Darwin B., and John P. Workman. 1971. The Importance of Renewable Grazing Resources on Federal Lands in the 11 Western States. Utah Agric. Expt. Sta. Cir. 155. 44 p.

Importance of federal land grazing to the livestock industry, and also the contribution such grazing makes to local communities and state economies.

Payne, Gene F. 1973. Vegetative Rangeland Types in Montana. Mon. Agric. Expt. Sta. Bul. 671. 16 p.

Classification and description of 22 rangeland types in Montana; rangeland type map of Montana included.

Reeves, Robert G. (Ed.). 1975. Manual of Remote Sensing, 2 Vols. Amer. Soc. Photogrammetry, Falls Church, Va. 2144 p.

A technical manual of remote sensing.

Sampson, Arthur W., Agnes Chase, and Donald W. Hedrick. 1951. California Grasslands and Range Forage Grasses. Calif. Agric. Expt. Sta. Bul. 724. 131 p.

Description, evaluation, and management of California grasslands; description, distribution, habitat, and forage value of the important native and naturalized range forage grasses; checklist of California grasses; plant name index; illustrated.

Sampson, Arthur W., and Beryl S. Jespersen. 1963. California Range Brushlands and Browse Plants. Calif. Agric. Expt. Sta. Manual 33. 162 p.

Description, evaluation and management of California range brushlands; description, distribution, economic value, and browse rating of the principal California browse plants; plant name index; illustrated.

Shaver, J.C. 1977. North Dakota Rangeland Resources, 1977. Soc. Range Mgt./Old West Regional Range Program, Denver, Colo. 118 p.

Physical characteristics, status, and productivity of North Dakota rangelands; range wildlife resources, range sites, and natural areas comprise appendix sections.

Shreve, Forrest. 1951. The Vegetation of the Sonoran Desert. Carnegie Inst. Wash. Pub. 591. 192 p.

An ecological treatise on the physical features and the vegetation of the Sonran desert of southwestern U.S. and Mexico.

Sneva, Forrest A., and Donald N. Hyder. 1962. Forecasting Range Herbage Production in Eastern Oregon. Oregon Agric. Expt. Sta. Bul. 588. 11 p.

Explanation and adaptation to ranges in eastern Oregon and adjacent areas of a technique for estimating normal range production and for forecasting range productivity in individual years.

Tueller, Paul T., Garwin Lorain, Karl Kipping, and Charles Wilkie. 1972. Methods for Measuring Vegetation Changes on Nevada Rangelands. Nev. Agric. Expt. Sta. Tech. Bul. 16. 55 p.

Comparison of using frequency sampling, ground photography, and aerial photography in measuring range trend.

Univ. of Idaho and Pacific Consultants. 1970. The Forage Resource, 4 Vols. Public Land Law Review Comm., Washington, D.C. Not consecutively paged.

Prepared to provide data on the forage resources of public lands including forage policies, nature and importance of the resources, and present and projected future uses.

USDA, Cons. Needs Inv. Comm. 1971. Basic Statistics--National Inventory of Soil and Water Conservation Needs, 1967. USDA Stat. Bul. 461. 211 p.

Tabular presentation of land uses and the conservation treatments needed for cropland, pastureland, and rangeland, including watersheds; 1967 used as base year.

USDA, For. Serv. 1977. National Forest Landscape Management, Volume 2, Chapter 3. Range. USDA Agric. Handbook 484. 43 p.

Application of landscape management concepts and principles to the visual aspects of range resources management; emphasis given to range vegetation control and range structures; fully illustrated.

USDA, Soil Cons. Serv. 1976. National Range Handbook. USDA, Soil Cons. Serv. Not paged.

Study, inventory, analysis, treatment, and management of the natural resources comprising native grazing land ecosystems; presents Soil Conservation Service policy and procedures for assisting in planning and applying rangeland resource conservation programs.

USDA, Soil Cons. Serv., Soil Survey Staff. 1951. Soil Survey Manual. USDA Agric. Handbook 18. 503 p.

Former system of soil classification used in the United States along with instructions for making and interpreting basic soil surveys.

_____. 1975. Soil Taxonomy: A Basic System of Soil Classification for Making and Interpreting Soil Surveys. USDA Agric. Handbook 436. 754 p.

Describes in detail the present system of soil classification used in the United States.

USDI, Bur. Land Mgt. 1973. Manual of Instructions for Survey of Public Lands of United States, 1973. USDI, Bur. Land Mgt. Tech. Bul. 6. 333 p.

Basic source of information for understanding and using completed surveys of public domain lands; rectangular system of federal land survey.

Weaver, J.E. 1954. North American Prairie. Johnsen Pub. Co., Lincoln, Neb. 348 p.

A semitechnical presentation of the ecology and management of the North American prairie.

Weaver, J.E., and F.W. Albertson. 1956. Grasslands of the Great Plains: Their Nature and Use. Johnsen Pub. Co., Lincoln, Neb. 395 p.

A semitechnical presentation of the ecology and management of the grasslands of the Great Plains.

West Reg. Soil Survey Work Group. 1964. Soils of the Western United States. West. Land Grant Univ. and Colleges and USDA, Soil Cons. Serv. 69 p.

Occurrence and distribution of soils in the 11 western states; describes 36 great soil groups and eight miscellaneous land types and discusses their relationships to physiography, climate, and vegetation; generalized soil association map.

Grazing Management

Australian Rangeland Soc. 1977. The Impact of Herbivores on Arid and Semiarid Rangelands. Proceedings of the 2nd United States/Australia Panel, Adelaide, 1972. Australian Rangeland Soc., Perth, W. Austr.

Papers covering the results of grazing impact research in the U.S. and Australia.

Beetle, A.A., W.M. Johnson, R.L. Lang, Morton May, and D.R. Smith. 1961. Effect of Grazing Intensity on Cattle Weights and Vegetation of the Bighorn Experimental Pastures. Wyo. Agric. Expt. Sta. Bul. 373. 23 p.

Results of a study on mountain range of northern Wyoming.

Bentley, J.R., and M.W. Talbot. 1951. Efficient Use of Annual Plants on Cattle Ranges in the California Foothills. USDA Cir. 870. 52 p.

Report of studies of grazing use of foothill range by cattle at the San Joaquin Experimental Range; emphasis given to adjusting ranch operations and stocking rates to fluctuating forage production.

Bohnning, J.W., and A.P. Thatcher. 1972. Proper Use and Management of Grazing Land. Ariz. Inter-Agency Range Comm., Tucson. 48 p.

Objectives, results, and procedures for proper usage of Arizona rangelands.

Boykin, C.C., J.R. Gray, and D.D. Caton. 1962. Ranch Production Adjustments to Drought in Eastern New Mexico. N. Mex. Agric. Expt. Sta. Bul. 470. 41 p.

A study of the effects of drought on ranch operation, costs, and income and an evaluation of ranching systems and adjustments to minimize drought effects.

Burzlaff, Donald F., and Lionel Harris. 1969. Yearling Steer Gains and Vegetation Changes in Western Nebraska Rangeland Under Three Rates of Stocking. Neb. Agric. Expt. Sta. Bul. 505. 18 p.

Results of a 10-year intensity of grazing study on sandy rangeland near Scottsbluff, Nebraska.

Cable, Dwight R. 1975. Range Management in the Chaparral Type and Its Ecological Basis: The Status of Our Knowledge. USDA, For. Serv. Res. Paper RM-155. 30 p.

Major sections on ecology and range management practices; good literature cited section; a reference for resource managers.

Cable, Dwight R., and S. Clark Martin. 1975. Vegetation Responses to Grazing, Rainfall, Site Condition, and Mesquite Control on Semidesert Range. USDA, For. Serv. Res. Paper RM-149. 24 p.

Report of vegetation changes from 1957 to 1966 on semidesert rangelands of the Santa Rita Experimental Range near Tucson.

Clary, Warren P. 1975. Range Management and Its Ecological Basis in the Ponderosa Pine Type of Arizona: The Status of Our Knowledge. USDA, For. Serv. Res. Paper RM-158. 35 p.

Emphasis given to physical-biological characteristics of the type, grazing management, and correlation of grazing with other uses; good literature cited section; a reference for resource managers.

Cook, C. Wayne. 1971. Effects of Season and Intensity of Use on Desert Vegetation. Utah Agric. Expt. Sta. Bul. 483. 57 p.

Results of herbage removal on the phenological and chemical response of major Utah desert species; also rate of recovery of plants in low vigor.

Currie, Pat O. 1975. Grazing Management of Ponderosa Pine-bunchgrass Ranges of the Central Rocky Mountains: The Status of Our Knowledge. USDA, For. Serv. Res. Paper RM-159. 24 p.

Grazing, its relationship to other uses, and further research needs in grazing management; good literature cited section; a reference for resource managers.

Currie, Pat O., and Dwight R. Smith. 1970. Response of Seeded Ranges to Different Grazing Intensities in the Ponderosa Pine Zone of Colorado. USDA Prod. Res. Rep. 112. 41 p.

Results of grazing experiments at the Manitou Experimental Forest on cool-season seeded pastures.

Frischknecht, Neil C., and Lorin E. Harris. 1968. Grazing Intensities and Systems on Crested Wheatgrass in Central Utah: Response of Vegetation and Cattle. USDA Tech. Bul. 1388. 47 p.

Results of a 14-year study at the Benmore Experimental Area, and recommendations for grazing management on similar ranges.

Gibbens, R.P., and H.G. Fisser. 1975. Influence of Grazing Management Systems on Vegetation in the Red Desert Region of Wyoming. Wyo. Agric. Expt. Sta. Sci. Monogr. 29. 23 p.

Applied research on the application of grazing systems to management situations.

Hedrick, D.W., J.A. Young, J.A.B. McArthur, and R.F. Keniston. 1968. Effects of Forest and Grazing Practices on Mixed Coniferous Forests of Northeastern Oregon. Ore. Agric. Expt. Sta. Tech. Bul. 103. 24 p.

Results of a study of forage production under different amounts of forest overstory and on logged versus unlogged areas, also methods of improving forage utilization; application of results to grazing management.

Herbel, Carlton H., Robert Steger, and Walter L. Gould. 1974. Managing Semidesert Ranges of the Southwest. N. Mex. Agric. Ext. Cir. 456. 48 p.

Recommendations from research and experience in the management of semiarid ranges, west Texas through Arizona; semitechnical publication for ranchers and range managers.

Holscher, Clark E., and E.J. Woolfolk. 1953. Forage Utilization by Cattle on Northern Great Plains Ranges. USDA Cir. 918. 27 p.

Study of factors affecting degree of use, and the improvement of grazing utilization.

Hormay, A.L., and M.W. Talbot. 1961. Rest-rotation Grazing--A New Management System for Perennial Bunchgrass Ranges. USDA Prod. Res. Rep. 51. 43 p.

Preliminary evaluation of rest-rotation grazing at Harvey Valley in northeastern California; recommendations for designing rest-rotation systems.

Houston, Walter R., and R.R. Woodward. 1966. Effects of Stocking Rates on Range Vegetation and Beef Cattle Production in the Northern Great Plains. USDA Tech. Bul. 1357. 58 p.

Results of an eight-year study of stocking rates at the U.S. Range Livestock Experiment Station near Miles City.

Hurtt, Leon C. 1951. Managing Northern Great Plains Cattle Ranges to Minimize Effects of Drought. USDA Cir. 865. 24 p.

Drought effects on range vegetation, cattle production, and ranch income with management recommendations to reduce drought effects.

Hutchings, Selar S., and George Stewart. 1953. Increasing Forage Yields and Sheep Production on Intermountain Winter Ranges. USDA Cir. 925. 63 p.

Results of a 13-year study of forage production, utilization, and grazing intensity at the Desert Experimental Range in western Utah; implications to the management of grazing on winter sheep ranges.

Hyder, D.N., R.E. Bement, E.E. Remmenga, and D.F. Hervey. 1975. Ecological Responses of Native Plants and Guidelines for Management of Shortgrass Range. USDA Tech. Bul. 1503. 87 p.

> Results of seasonal heavy grazing and nitrogen fertilization on perennial plant species, plant communities, and cattle at the Central Plains Experimental Range near Nunn, Colo.; applications to management of shortgrass range.

Johnson, W.M. 1953. Effect of Grazing Intensity Upon Vegetation and Cattle Gains on Ponderosa Pine-bunchgrass Ranges of the Front Range of Colorado. USDA Cir. 929. 36 p.

> Results of a grazing intensity study at the Manitou Experimental Forest near Colorado Springs.

_____. 1962. Vegetation of High-Altitude Ranges in Wyoming as Related to Use by Game and Domestic Sheep. Wyo. Agric. Expt. Sta. Bul. 387. 31 p.

> Herbage utilization by game and domestic sheep as related to plant species, plant communities, and other site characteristics.

Klipple, Graydon E. 1964. Early- and Late-season Grazing Versus Season-long Grazing of Short-grass Vegetation on the Central Great Plains. USDA, For. Serv. Res. Paper RM-11. 16 p.

> Results of a 10-year study of restricted grazing seasons on monthly cattle gains and herbage composition and production.

Klipple, G[raydon].E., and David F. Costello. 1960. Vegetation and Cattle Responses to Different Intensities of Grazing on Short-grass Ranges on the Central Great Plains. USDA Tech. Bul. 1216. 82 p.

> Fourteen-year study of grazing intensities at the Central Plains Experimental Range near Nunn, Colorado.

Kothmann, M.M., G.W. Mathis, P.T. Marion, and W.J. Waldrip. 1970. Livestock Production and Economic Returns from Grazing Treatments on the Texas Experimental Ranch. Texas Agric. Expt. Sta. Bul. 1100. 39 p.

> Results of a nine-year study on the effects of stocking rates, grazing systems, and supplemental winter feeding on cow-calf production in the Rolling Plains region of Texas.

Launchbaugh, J.L. 1957. The Effect of Stocking Rate on Cattle Gains and on Native Shortgrass Vegetation in West-central Kansas. Kan. Agric. Expt. Sta. Bul. 394. 29 p.

> Study includes the interrelationships among winter level of nutrition, summer stocking rates, and late summer supplementation of yearling steers; results of a ten-year study.

Lodge, Robert W., Sylvester Smoliak, and Alexander Johnston. 1972. Managing Crested Wheatgrass Pastures. Can. Dept. Agric. Pub. 1473. 20 p.

Establishment, growth, production, use, management, and renovation of crested wheatgrass pastures for livestock production.

Martin, S. Clark. 1975. Ecology and Management of Southwestern Semidesert Grass-shrub Ranges: The Status of Our Knowledge. USDA, For. Serv. Res. Paper RM-156. 39 p.

Application of research and experience to the management of this range type extending from southwestern Texas through Arizona; prepared for ranchers and range administrators; ample literature citations.

______. 1975. Stocking Strategies and Net Cattle Sales on Semidesert Range. USDA, For. Serv. Res. Paper RM-146. 10 p.

Study of methods of coping with year-to-year changes in range forage production and probable effects of these adjustments on ranch income.

Martin, S. Clark, and Charles R. Whitfield. 1973. Grazing Systems for Arizona Ranges. Ariz. Interagency Range Comm., Tucson. 36 p.

Includes guidelines for the successful development and application of grazing systems.

Martin, S. Clark, and Dwight R. Cable. 1974. Managing Semidesert Grass-shrub Ranges. USDA Tech. Bul. 1480. 45 p.

Study to evaluate factors casuing vegetational changes on semidesert grass-shrub cattle ranges; study conducted on the Santa Rita Experimental Range near Tucson.

Paulsen, Harold A., Jr. 1975. Range Management in the Central and Southern Rocky Mountains: A Summary of the Status of Our Knowledge by Range Ecosystems. USDA, For. Serv. Res. Paper RM-154. 34 p.

Summarization of a series of comprehensive reports on ecology and management of seven recognized range ecosystems in the central and southern Rocky Mountains.

Paulsen, Harold A., Jr., and Fred N. Ares. 1962. Grazing Values and Management of Black Grama and Tobosa Grasslands and Associated Shrub Ranges of the Southwest. USDA Tech. Bul. 1270. 56 p.

Results of studies on the Jornada Experimental Range near Las Cruces to determine the characteristics of the major forage species and vegetation types and methods of management best suited to the region.

Pechanec, Joseph F., and George Stewart. 1949. Grazing Spring-fall Sheep Ranges of Southern Idaho. USDA Cir. 808. 34 p.

Recommendations based on research and experience for evaluating range productivity, planning grazing programs, and making periodic adjustments in grazing management.

Pickford, G.D., and Elbert H. Reid. 1948. Forage Utilization on Summer Cattle Ranges in Eastern Oregon. USDA Cir. 796. 27 p.

Results of a study at the Starkey Experimental Range.

Pieper, Rex D. 1970. Species Utilization and Botanical Composition of Cattle Diets on Pinyon-juniper Grassland. N. Mex. Agric. Expt. Sta. Bul. 566. 16 p.

Results of a study on the Fort Stanton station in central New Mexico on grazing patterns, species utilization, and seasonal cattle diets.

Pond, Floyd W., and Dixie R. Smith. 1971. Ecology and Management of Subalpine Ranges on the Big Horn Mountains of Wyoming. Wyo. Agric. Expt. Sta. Res. J. 53. 25 p.

Summarization of studies on plant communities, plant growth and development, grazing practices, and range development.

Ratliff, Raymond D., Jack N. Reppert, and Richard J. McConnen. 1972. Rest-rotation Grazing at Harvey Valley: Range Health, Cattle Gains, Costs. USDA, For. Serv. Res. Paper PSW-77. 24 p.

An interim report of rest-rotation grazing at Harvey Valley in northeastern California.

Rauzi, Frank, and Robert L. Lang. 1967. Effect of Grazing Intensity on Vegetation and Sheep Gains on Shortgrass Rangeland. Wyo. Agric. Expt. Sta. Sci. Monogr. 4. 11 p.

Results of a 10-year study at the Archer Station in southeastern Wyoming.

Reed, Merton J., and Roald A. Peterson. 1961. Vegetation, Soil, and Cattle Responses to Grazing on Northern Great Plains Range. USDA Tech. Bul. 1252. 79 p.

Results of a 14-year study of grazing intensities at the U.S. Range Livestock Expt. Sta. near Miles City, Mon.

Reynolds, Hudson G., and S. Clark Martin. 1968 (Rev.). Managing Grass-shrub Cattle Ranges in the Southwest. USDA Agric. Handbook 162. 44 p.

Recommendations for management of the grass-shrub type based on 40 years of intensive work on the Santa Rita Experimental Range near Tucson.

Sarvis, J.T. 1941. Grazing Investigations on the Northern Great Plains. N. Dak. Agric. Expt. Sta. Bul. 308. 110 p.

Results of 25 years of research on intensities and methods of grazing on native range and cattle near Mandan, N. Dak.

Sharp, Lee A. 1970. Suggested Management Programs for Grazing Crested Wheatgrass. Idaho For., Wildl., and Range Expt. Sta. Bul. 4. 19 p.

Results of grazing management research on crested wheatgrass range in southern Idaho and suggested management programs for similar ranges.

Shiflet, Thomas N., and Harold F. Heady. 1971. Specialized Grazing Systems--Their Place in Range Management. USDA, SCS TP-152. 13 p.

A review of the published studies on specialized grazing systems--their advantages, disadvantages, and uses.

Sims, Phillip L., B.E. Dahl, and A.H. Denham. 1976. Vegetation and Livestock Response at Three Grazing Intensities on Sandhill Rangeland in Eastern Colorado. Colo. Agric. Expt. Sta. Tech. Bul. 130. 48 p.

Results of a 12-year study at the Eastern Colorado Range Station near Akron.

Skovlin, Jon M. 1965. Improving Cattle Distribution on Western Mountain Rangelands. USDA Farm. Bul. 2212. 14 p.

Problems of grazing distribution and recommendations for improving cattle distribution with emphasis on western mountain rangelands.

Skovlin, Jon M., Robert W. Harris, Gerald S. Strickler, and George A. Garrison. 1976. Effects of Cattle Grazing Methods on Ponderosa Pine-bunchgrass Range in the Pacific Northwest. USDA Tech. Bul. 1531. 40 p.

Results of a 11-year study of stocking rates and grazing systems on the Starkey Experimental Forest and Range and the effects of grazing treatments on other natural resource uses.

Smith, Dwight R. 1967. Effects of Cattle Grazing on a Ponderosa Pine-bunchgrass Range in Colorado. USDA Tech. Bul. 1371. 60 p.

Relationships among three intensities of grazing on range utilization, herbage production, other plant resources, soil conditions, and cattle weight gains.

Smoliak, S., A. Johnston, M.R. Kilcher, and R.W. Lodge. 1976. Management of Prairie Rangeland. Can. Dept. Agric. Pub. 1589. 30 p.

The application of range and pasture studies to the grasslands and adjacent parklands of western Canada.

Smoliak, S., A. Johnston, R.A. Wrote, and M.G. Turnbull. 1975. Alberta Range Pastures. Alberta Dept. Agric. Pub. 134/14. 29 p.

Description of Alberta range types, range condition rating, and stocking rates.

Springfield, H.W. 1976. Characteristics and Management of Southwestern Pinyon-juniper Ranges: The Status of Our Knowledge. USDA, For. Serv. Res. Paper RM-160. 32 p.

Modification and management of southwestern pinyon-juniper ranges to provide an optimum mix of forage for livestock, food and habitat for wild animals, and other products; recommendations for ranchers and range administrators.

Theurer, C. Brent, A.L. Lesperance, Joe D. Wallace, A.H. Denham, D.C. Clanton, D.A. Price, and L.E. Harris. 1976. Botanical Composition of the Diet of Livestock Grazing Native Ranges. Ariz. Agric. Expt. Sta. Tech. Bul. 233. 19 p.

Reviews and evaluates methods of botanical analyses of livestock diets; presents results of dietary studies with cattle and sheep grazing native ranges in the western United States.

Thilenius, John F. 1975. Alpine Range Management in the Western United States--Principles, Practice, and Problems: The Status of Our Knowledge. USDA, For. Serv. Res. Paper RM-157. 32 p.

Characteristics and ecology of high-elevation, cold-dominated, alpine ecosystems; management of these ecosystems, with emphasis on the use of the forage resource and its relationship with other uses.

Turner, George T., and Harold A. Paulsen, Jr. 1976. Management of Mountain Grasslands in the Central Rockies: The Status of Our Knowledge. USDA, For. Serv. Res. Paper RM-161. 24 p.

Proper grazing management, improving forage production, and their relationships with other uses; written for the rancher and range administrator.

Wagnon, Kenneth A. 1963. Behavior of Beef Cows on a California Range. Calif. Agric. Expt. Sta. Bul. 799. 58 p.

Study of effects of natural range conditions, stocking rates, and supplementation on range cattle behavior at the San Joaquin Experimental Range.

_____. 1968. Use of Different Classes of Range Land by Cattle. Calif. Agric. Expt. Sta. Bul. 838. 16 p.

Differential grazing of topographic sites in the California annual-type foothill range and the effects of natural and management factors.

Woolfolk, E.J. 1949. Stocking Northern Great Plains Sheep Range for Sustained High Production. USDA Cir. 804. 39 p.

Plant growth and development, habits of yearling ewes, and stocking rates interrelationships; grazing recommendations based on research at the U.S. Range Livestock Experiment Station near Miles City.

_____. 1949. Weight and Gain of Range Calves as Affected by Rate of Stocking. Mon. Agric. Expt. Sta. Bul. 463. 26 p.

Results of research at the Range Livestock Experiment Station near Miles City.

Range and Ranch Development

Alley, H.P., A.F. Gale, and N.E. Humburg. 1978. Wyoming Weed Control Guide, 1978. Wyo. Agric. Ext. Bul. 442. 53 p.

Cultural and chemical recommendations for controlling weeds in Wyoming; includes sections on perennial weeds, poisonous plants, range and pasture, and woody plants.

Ariz. Interagency Range Tech. Comm. 1969. Guide to Improvement of Arizona Rangeland. Ariz. Agric. Ext. Bul. A-58. 93 p.

Provides guidelines for planning and application of proper range use, brush control, range seeding, and fertilization to Arizona rangeland; also includes descriptions of the environmental zones (vegetation types) of the state.

Artz, John L., J. Boyd Price, Frederick F. Peterson, Harry B. Summerfield, Richard E. Eckert, et al. 1970. Plantings for Wildlands and Erosion Control. Nev. Agric. Ext. Cir. 108. 24 p.

Recommendations on rangeland forage plantings, wildlife plantings, soil stabilization plantings, and windbreak plantings for Nevada.

Barnes, Oscar K., and Darwin Anderson. 1958. Pitting for Range Improvement in the Great Plains and the Southwest Desert Regions. USDA Prod. Res. Rep. 23. 17 p.

Summarizes studies and experiences in range pitting with specific reference to northern shortgrass plains, southern shortgrass plains, and the southwest desert area.

Bentley, Jay R. 1967. Conversion of Chaparral Areas to Grassland: Techniques Used in California. USDA Agric. Handbook 328. 35 p.

Considers evaluation of benefits, selection of the area, removal of brush, establishment of a grass cover, and control of regrowth as it pertains to conversion of California chaparral areas.

Bovey, Rodney W. 1977. Response of Selected Woody Plants in the United States to Herbicides. USDA Agric. Handbook 493. 101 p.

Summarization in tabular form of the response of brush species and trees to the more promising herbicides; also includes information on application methods, characteristics of herbicides, herbicide formulations, and care of equipment.

Calif. Fertilizer Assoc., Soil Improvement Comm. 1975. Western Fertilizer Handbook. Interstate Print. and Pub., Danville, Ill. 242 p.

Fertilization and recommendations for California; includes pastures, ranges, and harvested forages.

Conard, E.C. 1962. How to Establish New Pastures. Neb. Agric. Ext. Campaign Cir. 165. 12 p.

Recommendations on planning, site preparation, and seeding of perennial range and pasture in Nebraska.

Cook, C. Wayne. 1958. Sagebrush Eradication and Broadcast Seeding. Utah Agric. Expt. Sta. Bul. 404. 23 p.

Research results of studies in central Utah on methods of sagebrush removal, season of seeding, methods of broadcast seeding, and adaptation of introduced wheatgrass species.

_____. 1965. Plant and Livestock Responses to Fertilized Rangelands. Utah Agric. Expt. Sta. Bul. 455. 35 p.

Results of Utah research on fertilizing seeded foothill and mountain range and native mountain range and the use of starter fertilizer on semiarid and mountain range seedings.

_____. 1966. Development and Use of Foothill Ranges in Utah. Utah Agric. Expt. Sta. Bul. 461. 47 p.

Results of research on developing spring-use foothill ranges in central Utah and management and livestock response from grazing seeded areas.

Cordingly, Robert V., and W. Gordon Kearl. 1975. Economics of Range Reseeding in the Plains of Wyoming. Wyo. Agric. Expt. Sta. Res. J. 98. 39 p.

Study to determine the costs of various range reseeding practices, determine their physical results, and determine returns and use of reseeded rangeland.

Cords, H.P., and J.L. Artz. 1976 (Rev.). Rangeland, Irrigated Pasture, and Meadow Weed Control Recommendations. Nev. Agric. Ext. Cir. 148. 3 p.

Practical, recommended weed control for grazing lands and hay meadows.

Cornelius, Donald R., and M.W. Talbot. 1955. Rangeland Improvement through Seeding and Weed Control on East Slope Sierra Nevada and on Southern Cascade Mountains. USDA Agric. Handbook 88. 51 p.

Results of research on sagebrush sites, dry grassland sites, weedy mountain meadows, and pine, aspen, and red fir types.

Dimeo, Art. 1977. An Investigation of Equipment for Rejuvenating Browse. USDA, For. Serv., Equip. Dev. Center (Missoula, Mon.). ED&T 7080. 19 p.

Preliminary evaluation of mechanical and hand equipment for rejuvenating browse plants.

Duran, Gilbert, and H.F. Kaiser. 1972. Range Management Practices: Investment Costs, 1970. USDA Agric. Handbook 435. 38 p.

Investment cost data utilized in the Forest-range Environmental Study (FRES) for making range improvements using 1970 as base year.

Eckert, Richard E., Jr. 1975. Improvement of Mountain Meadows in Nevada. USDI, Bur. Land Mgt., Washington, D.C. 45 p.

Results of research on seeding studies, iris control, fertilization, and tree and shrub transplants.

Elwell, Harry M., and W.E. McMurphy. 1973. Weed Control With Phenoxy Herbicides on Native Grasslands. Okla. Agric. Expt. Sta. Bul. 706. 23 p.

Results of several experiments on herbicidal control of weeds in native Oklahoma grasslands.

Ensign, R.D., and H.L. Harris. (Eds.). 1975. Idaho Forage Crop Handbook. Idaho Agric. Expt. Sta. Bul. 547. 54 p.

Recommendations for establishment, management, and utilization of grasses and legumes for hay, pasture (including dryland range), and silage in Idaho.

Frasier, Gary W. 1975. Proceedings of the Water Harvesting Symposium, Phoenix, Arizona, March 26-28. 1974. USDA ARS W-22. 329 p.

Papers on water harvesting in the development of local water resources for livestock, wildlife, runoff farming, and domestic use.

Friesen, H.A., M. Aaston, W.G. Corns, J.L. Dobb, and A. Johnston. 1965. Brush Control in Western Canada. Can. Dept. Agric. Pub. 1240. 28 p.

Chemical and mechanical brush control and seeding in woody vegetation types for range improvement in western Canada.

Gifford, Gerald F., and Frank E. Busby. (Eds.). 1975. The Pinyon-juniper Ecosystem: A Symposium, May, 1975. Utah State Univ., Logan, Utah. 194 p.

Characteristics of pinyon-juniper ecosystems, methods and benefits of woody plant control, impact of ecosystem manipulation, and management strategies.

Gomm, F.B. 1974. Forage Species for the Northern Intermountain Region: A Summary of Seeding Trials. USDA Tech. Bul. 1479. 307 p.

Summarizations of many range seeding trials made on diverse sites throughout Montana to determine the adaptability and forage value of forage species and strains.

Gray, James R., Thomas M. Stubblefield, and N. Keith Roberts. 1965. Economic Aspects of Range Improvements in the Southwest. N. Mex. Agric. Expt. Sta. Bul. 498. 47 p.

Costs of range improvements in the Southwest and returns necessary to pay for these improvements; a planning guide for range improvements.

Green, Lisle R. 1977. Fuelbreaks and Other Fuel Modification for Wildland Fire Control. USDA Agric. Handbook 499. 79 p.

Methods of preparing fuelbreaks and their use in fire management, with particular emphasis on California chaparral.

Hafenrichter, A.L., John L. Schwendiman, Harold L. Harris, Robert S. MacLauchlan, and Harold D. Miller. 1968. Grasses and Legumes for Soil Conservation in the Pacific Northwest and Great Basin States. USDA Agric. Handbook 339. 69 p.

Descriptions (with illustrations) and conservation uses of grass and legume species and varieties tested for northwestern U.S. by SCS; includes agricultural zone maps and descriptive legends for states of Washington, Oregon, California, Idaho, Nevada, and Utah.

Halls, L.K., R.H. Hughes, and F.A. Peevy. 1960. Grazed Firebreaks in Southern Forests. USDA Agric. Info. Bul. 226. 8 p.

A guide for establishing and maintaining grazed firebreaks, i.e., fuelbreaks, in Southern forests.

Hanson, A.A. 1965 (Rev.). Grass Varieties in the United States. USDA Agric. Handbook 170. 102 p.

A reference manual on the origin and current status of named and experimental grass varieties, including information on source material, varietal characteristics, and seed supplies.

Harris, Harold L., A.E. Slinkard, and A.L. Hafenrichter. 1972. Establishment and Production of Grasses Under Semiarid Conditions in the Intermountain West. Idaho Agric. Expt. Sta. Bul. 532. 16 p.

Results of semiarid site seeding studies in Idaho with emphasis on land preparation methods, methods of seeding, adapted species and varieties, and mixtures.

Herbel, Carlton H., and Walter L. Gould. 1973. Improving Arid Rangelands. Jornada Expt. Range Rep. 4. 16 p.

Recommendations for maintaining and improving arid rangelands of the Southwest through brush control, seeding, and improved management based on research at the Jornada Experimental Range.

Horvath, Joseph, Dennis Schweitzer, and Enoch Bell. 1978. Grazing on National Forest System Lands: Cost of Increasing Capacity in the Northern Region. USDA, For. Serv. Res. Paper INT-215. 56 p.

A survey of the potential grazing capacity and cost of additional range improvements and the value of existing improvements.

Houston, Walter R. 1971. Range Improvement Methods and Environmental Influences in the Norther Great Plains. USDA Prod. Res. Rep. 130. 13 p.

Comparison of land treatments for range improvement including protection from grazing, seeding, pitting, and nitrogen fertilization near Miles City, Mon.

Hubbard, William A. 1975 (Rev.). Farm and Ranch Equipment for Beef Cattle. Can. Dept. Agric. Pub. 1390. 37 p.

Suggestions with diagrams for construction of efficient beef cattle equipment for farm and ranch use.

Hull, A.C., Jr., D.F. Hervey, Clyde W. Doran, and W.J. McGinnies. 1958. Seeding Colorado Range Lands. Colo. Agric. Expt. Sta. Bul. 498. 46 p.

Recommendations for seeding rangelands of Colorado; emphasis given to site preparation, planting, species adaptation, and management of seeded stands.

Hull, A.C., Jr., and Ralph C. Holmgren. 1964. Seeding Southern Idaho Rangelands. USDA, For. Serv. Res. Paper INT-10. 31 p.

Summarization of available information on range seeding, with recommendations primarily being based on experimental seedings.

Hull, A.C., Jr., and W.M. Johnson. 1955. Range Seeding in the Ponderosa Pine Zone in Colorado. USDA Cir. 953. 40 p.

Summarization of research results and experience for use as a guide to successful range seeding in Colorado's ponderosa pine zone.

Jensen, Louis A., John O. Evans, J. LaMar Anderson, and Alvin R. Hamson. 1978 (Rev.). Chemical Control Guide, Utah, 1977 (and 1978 Addendum). Utah Agric. Ext. Cir. 301. 58 and 7 p.

Herbicidal plant control recommendations for Utah.

Klingman, D.L., and W.C. Shaw. 1975 (Rev.). Using Phenoxy Herbicides Effectively. USDA Farm. Bul. 2183. 25 p.

Formulation, application, and safety in the use of phenoxy herbicides; includes tables of susceptibility of common weeds to 2,4-D, MCPA, 2,4,5-T, silvex, and 2,4-DB.

Lamb, Samuel H., and Rex Pieper. 1971. Game Range Improvement in New Mexico. Ariz. Interagency Range Comm. Rep. 9. 28 p.

Game range improvement recommendations by vegetative types and by game species for New Mexico.

Lang, Robert, Frank Rauzi, Wesley Seamands, and Gene Howard. 1975. Guidelines for Seeding Range, Pasture, and Disturbed Lands. Wyo. Agric. Expt. Sta. Bul. 621. 11 p.

A guide for range seeding in Wyoming; includes general seeding recommendations and specific recommendations for respective precipitation zones.

Larson, Russell E., and Richard O. Hegg. 1976 (Rev.). Feedlot and Ranch Equipment for Beef Cattle. USDA Farm. Bul. 1584. 20 p.

Popular explanation of efficient beef cattle equipment for ranch and feedlot use.

Lauritzen, C.W., and Arnold A. Thayer. 1966. Rain Traps for Intercepting and Storing Water for Livestock. USDA Agric. Info. Bul. 307. 10 p.

Use, location, site preparation, design, and construction of rain traps for livestock water.

Lawrence, T., and D.H. Heinrichs. 1977. Growing Russian Wild Ryegrass in Western Canada. Can. Dept. Agric. Pub. 1607. 27 p.

Adaptation, available cultivars, and use and management recommendations for Russian wildrye for pasture, hay, farmyards, and seed production.

Lloyd, Russell D., and C. Wayne Cook. 1960. Seeding Utah's Ranges--An Economic Guide. Utah Agric. Expt. Sta. Bul. 423. 19 p.

An evaluation of costs and returns of crested wheatgrass seedings made by the Bureau of Land Management on 54,000 acres in western Utah, 1952-4.

Lowman, Ben, Dan McKenzie, and Dick Hallman. 1974. Investigation of Selected Problems in Range Habitat Improvement. USDA, For. Serv. Equip. Dev. Centers. 45 p.

Results of a 1972 survey of equipment problems and needs for range improvement, with a status report on solving the problems most frequently encountered.

Martin, Robert E., and John D. Dell. 1978. Planning for Prescribed Burning in the Inland Northwest. USDA, For. Serv. Gen. Tech. Rep. PNW-76. 67 p.

The role of fire, its potential uses, and planning prescribed burns in forests and ranges of the inland Pacific Northwest.

McCorkle, C.O., Jr., and D.D. Caton. 1962. Economic Analysis of Range Improvements: A Guide for Western Ranchers. Calif. Agric. Expt. Sta. Giannini Foundation Res. Rep. 255. 79 p.

Manual of procedures for economic analysis of anticipated range improvements.

McLean, Alastair, and A[lfred].H. Bawtree. 1962. Seeding Grassland Ranges in the Interior of British Columbia. Can. Dept. Agric. Pub. 1444. 11 p.

Recommendations for seeding interior grassland ranges; includes appendix table on adaptation of common grasses and legumes.

______. 1973. Seeding Forest Rangelands in British Columbia. Can. Dept. Agric. Pub. 1463. 14 p.

Recommendations for seeding forest rangelands; includes appendix table on adaptation of common grasses and legumes.

McNamee, Michael A., and Edwin A. Kinne. 1965. Pasture and Range Fences. Rocky Mtn. Regional Publication 2. (Univ. Wyo., Laramie). 32 p.

Fencing materials, fence types, construction methods, and maintenance; practical recommendations, well illustrated; prepared by extension specialists in several western states.

Mayland, H.F. (Ed.). 1973. Water-animal Relations Symposium Proceedings. Water-Animal Relations Comm., Kimberly, Ida. 239 p.

A symposium of 20 papers on various phases of livestock water quality, water needs and factors affecting intake, grazing management, and specialized water developments.

Mays, D.A. (Ed.). 1974. Forage Fertilization. Amer. Soc. Agron., Madison, Wisc. 621 p.

Review papers on fertilization of rangeland, pastures, and harvested forage crops.

Natl. Res. Council, Subcomm. on Weeds. 1968. Principles of Plant and Animal Pest Control. Volume 2. Weed Control. Natl. Acad. Sci., Washington, D.C. (Pub. 1597). 471 p.

Basic problems, principles, and research needs in weed control.

Nielsen, Darwin B. 1967. Economics of Range Improvements. Utah Agric. Expt. Sta. Bul. 466. 49 p.

A guide for making economic decisions about range improvements; suggests steps and procedures for decision making; provides examples.

N. Mex. Interagency Range Comm. 1968. Improving Pinyon-juniper Ranges in New Mexico. N. Mex. Interagency Range Comm. Rep. 2. 23 p.

Interagency recommendations for improving pinyon-juniper ranges in New Mexico.

Olson, Carl E., William A. Daley, and Charles C. McAfee. 1977. An Economic Evaluation of Range Resource Improvement. Wyo. Agric. Expt. Sta. Bul. 650. 14 p.

Describes a method for determining the profitability of rangeland improvement with implications for internal ranch expansion and net ranch income.

Owensby, C.E., and J.L. Launchbaugh. 1976. Controlling Weeds and Brush on Range Land. Kan. Agric. Ext. Leaflet 429. 8 p.

Recommendations for controlling common weed and brush species on Kansas ranges.

Parker, Karl G. 1968. Range Plant Control Modernized. Utah Agric. Ext. Cir. 346. 13 p.

Recommendations for controlling undesirable plants on Utah rangelands.

Parker, Robert, and Lyle A. Derscheid. 1970. Chemical Weed Control in Pasture, Range and Hayland. S. Dak. Agric. Ext. Fact Sheet 426. 5 p.

Herbicidal plant control recommendations in South Dakota.

Pase, Charles P., and Carl Eric Granfelt (Eds.). 1977. The Use of Fire on Arizona Rangelands. Ariz. Interagency Range Comm. Pub. 4. 15 p.

Recommendations for using fire to accomplish desired manipulation of the major range plant communities in Arizona.

Pechanec, Joseph F., A. Perry Plummer, Joseph H. Robertson, and A.C. Hull, Jr. 1965. Sagebrush Control on Rangelands. USDA Agric. Handbook 277. 40 p.

Summarization of information on sagebrush control and management needed after control, with primary emphasis given to big sagebrush in the Intermountain region.

Plummer, A. Perry, A.C. Hull, Jr., George Stewart, and Joseph H. Robertson. 1955. Seeding Rangelands in Utah, Nevada, Southern Idaho and Western Wyoming. USDA Agric. Handbook 73. 71 p.

A handbook of range seeding recommendations for the Intermountain region.

Plummer, A. Perry, Donald R. Christensen, and Stephen B. Monsen. 1968. Restoring Big Game Range in Utah. Utah Div. Fish and Game Pub. 68-3.

Describes treatments and procedures, based on research and experience, for improving big game range in Utah; special emphasis given to revegetation with adapted plant species.

Price, J.D., Rupert D. Palmer, and G.O. Hoffman. 1974. (Rev.). Suggestions for Weed Control with Chemicals--Range and Timber Lands and Pasture and Forage Crops. Texas Agric. Ext. Misc. Pub. 1060. 16 p.

General and specific recommendations for chemical plant control in range, pasture, and forage crops in Texas.

Rauzi, Frank. 1968. Pitting and Interseeding Native Shortgrass Rangeland. Wyo. Agric. Expt. Sta. Res. J. 17. 14 p.

Results of studies of pitting and interseeding on range vegetation and sheep gains in southeastern Wyoming.

Robison, Laren, Dave Nuland, and John Furrer. 1970. Chemicals that Control Weeds. Neb. Agric. Ext. Cir. 70-130. 8 p.

Includes section on undesirable plant control in pastures, ranges, and forage crops; also section on troublesome weeds and woody plants.

Roby, George A., and Lisle R. Green. 1976. Mechanical Methods of Chaparral Modification. USDA Agric. Handbook 487. 46 p.

Describes and evaluates alternative techniques and equipment for brush crushing, compacting, chopping, and shredding and for grass seeding.

Ross, J.G., and C.R. Krueger. 1976. Grass Species and Variety Performance in South Dakota. S. Dak. Agric. Expt. Sta. Bul. 642. 47 p.

A report and synthesis of selected grass breeding and pasture projects in South Dakota.

Skinner, J., R. Rae, and R. Kapty. 1975. Farm Pasture Fencing. Can. Dept. Agric. Pub. 1568. 31 p.

Planning and constructing pasture and field fences for the control of livestock; includes illustrations; recommendations adapted to range fencing.

Spencer, E.Y. 1973 (6th Ed.). Guide to the Chemicals Used in Crop Protection. Can. Dept. Agric. Pub. 1093. 483 p.

Chemical structure, development and manufacture, properties, formulations, and analyses for chemicals used in crop protection, including herbicides and insecticides.

USDA and USDI, Range Seeding Equip. Comm. 1957 (Rev.). Handbook of Range Seeding Equipment. U.S. Govt. Printing Office, Washington, D.C. Various pagings.

Description of equipment adapted or designed for use in range seeding and noxious range plant control.

_____. 1966. Chemical Control of Range Weeds. U.S. Govt. Print. Office, Washington, D.C. 39 p.

Provides suggestions on the control of range weeds (including shrubs) with chemicals; prepared by western range research and management personnel.

USDA, Agric. Res. Serv. 1973. Guidelines for Weed Control. USDA Agric. Handbook 447. 169 + p.

Informational source on the use of chemicals in developing weed control recommendations.

_____. 1976 (Rev.). Aerial Application of Agricultural Chemicals. USDA Agric. Handbook 287. 25 p.

A handbook for training agricultural aircraft pilots and a reference guide for aerial applicators; includes useful information for range technicians.

USDA, For. Serv. 1969. Structural Range Improvement Handbook. U.S. For. Serv., Intermtn. Reg., Ogden, FSH 2209.22.

Materials, construction details, and maintenance needs for various types of range fences and range water developments.

_____. 1969. Wildlife Habitat Improvement Handbook. USDA, For. Serv., Washington, D.C. Not paged.

Includes major section on upland improvements (range-related) as well as sections on stream, lake, wetland, and special structural improvements.

Vallentine, John F. 1963. Water for Range Livestock. Neb. Agric. Ext. Cir. 63-156. 16 p.

Planning, development, and use of livestock watering facilities on range and pasture lands.

Vallentine, John F., C. Wayne Cook, and L.A. Stoddart. 1963. Range Seeding in Utah. Utah Agric. Ext. Cir. 307. 20 p.

Summarization of range seeding recommendations for livestock ranges in Utah.

Wambolt, Carl. 1976. Montana Range Seeding Guide. Mon. Agric. Ext. Bul. 347. 22 p.

Summarization of range seeding recommendations for Montana.

Wash. Agric. Ext. Serv. 1977 (Rev.). Weed Control in Rangelands, Pastures and Haylands. Wash. Agric. Ext. EM 3452. 11 p.

Emphasis given to chemical control of selected undesirable plants and plant groups.

Weed Sci. Soc. Amer., Handbook Comm. 1979. Herbicide Handbook of the Weed Science Society of America. Champaign, Ill. 479 p.

Nomenclature, chemical and physical properties, herbicidal use, use precautions, physiological and biochemical behavior, toxicological properties, and analytical methods for individual herbicides.

Wright, Henry A. 1978. The Effect of Fire on Vegetation in Ponderosa Pine Forests: A State-of-the-Art Review. Texas Tech. Univ. Range and Wildl. Info. Series 2. 21 p.

Emphasis given to the use and effects of fire, management implications, and research needs for six regions of the country where ponderosa pine grows.

Range Livestock

Albaugh, Reuben, and Horace T. Strong. 1972 (Rev.). Breeding Yearling Beef Heifers. Calif. Agric. Ext. Cir. 433. 21 p.

Management recommendations for breeding yearling beef heifers.

Allen, W.L. 1969. Sheep Raising in Canada. Can. Dept. of Agric. Pub. 1401. 55 p.

A popular manual on sheep production practices.

Anderson, M.S., H.W. Lakin, K.C. Beeson, Floyd F. Smith, and Edward Thacker. 1961. Selenium in Agriculture. USDA Agric. Handbook 200. 65 p.

Problems of selenium in soils, its absorption by plants, and toxic effects on animals consuming seleniferous plants.

Bellows, R.A., W.C. Foote, M.L. Hopwood, and P.T. Cupps (Eds.). 1971. Physiology of Reproduction in Cattle. Calif. Agric. Expt. Sta. Bul. 853. 66 p.

Summarization of western regional research on physiological factors which control normal reproduction in cattle.

Beverly, John R. 1975. Recognizing and Handling Calving Problems. Texas Agric. Expt. Sta. Misc. Pub. 1203. 8 p.

Practical recommendations on deciding when, what, and how to assist in difficult calving.

Blair, R.M., and E.A. Epps, Jr. 1969. Seasonal Distribution of Nutrients in Plants of Seven Browse Species in Louisiana. USDA, For. Serv. Res. Paper SO-51. 35 p.

Nutrient trends in selected browse species and the distribution of nutrients among plant parts.

Burzlaff, Donald F., W.F. Wedin, E.H. McIlvain, Warren C. Thompson, et al. 1976. Proceedings for Symposium on Integration of Resources for Beef Cattle Production, February 16-20, 1976, Omaha, Nebraska. Society for Range Management, Denver, Colo. 75 p.

Papers on complementary relationships of range, temporary pasture, and harvested forages in meeting beef cattle production needs.

Campbell, Robert S., E.A. Epps, Jr., C.C. Moreland, J.L. Farr, and Frances Bonner. 1954. Nutritive Values of Native Plants on Forest Range in Central Louisiana. La. Agric. Expt. Sta. Bul. 488. 18 p.

Chemical analyses of important range forage species in relation to stage of growth, management practices, and dietary deficiencies.

Can. Dept. Agric. 1976. Sheep Production and Marketing. Can. Dept. Agric. Pub. 1582. Various pagings.

A syllabus on the management, breeding, feeding, flock health, marketing, and buildings and equipment for practical sheep production.

Cardon, B.P., E.B. Stanley, W.J. Pistor, and J.C. Nesbit. 1951. Use of Salt as a Regulator of Supplemental Feed Intake and Its Effect on the Health of Range Livestock. Ariz. Agric. Expt. Sta. Bul. 239. 15 p.

Results of research into high salt tolerance of ruminants and the feasibility of salt as a regulator of supplemental feed intake by range livestock.

Cook, C. Wayne. 1956. Range Livestock Nutrition and Its Importance in the Intermountain Region. The Faculty Assoc., Utah State Agric. College, Logan. 28 p.

A faculty research lecture summarizing the knowledge of range livestock nutrition in the Intermountain region to 1956.

_____. 1970. Energy Budget of the Range and Range Livestock. Colo. Agric. Expt. Sta. Tech. Bul. 109. 28 p.

Develops a procedure for estimating the energy yield of the range by summarizing the energy required for all of the physiological functions performed by the grazing animals.

Cook, C. Wayne, L.A. Stoddart, and Lorin E. Harris. 1954. Nutritive Value of Winter Range Plants in the Great Basin. Utah Agric. Expt. Sta. Bul. 372. 56 p.

Results of studies on nutritive value of winter range plants in Utah and implications for supplementation.

_____. 1956. Comparative Nutritive Value and Palatability of Some Introduced and Native Forage Plants for Spring and Summer Grazing. Utah Agric. Expt. Sta. Bul. 385. 39 p.

Results of studies with range sheep in central Utah.

Cook, C. Wayne, and Lorin E. Harris. 1950. The Nutritive Value of Range Forage as Affected by Vegetation Type, Site, and Stage of Maturity. Utah Agric. Expt. Sta. Tech. Bul. 344. 45 p.

Effects of natural factors on the chemical composition of mountain range plants of northern Utah.

_____. 1950. Nutritive Content of the Grazing Sheep's Diet on Summer and Winter Ranges of Utah. Utah Agric. Expt. Sta. Bul. 342. 66 p.

Results of studies on factors affecting the nutrient content of the diets of grazing sheep.

______. 1968. Effect of Supplementation on Intake and Digestibility of Range Forage. Utah Agric. Expt. Sta. Bul. 475. 38 p.

Results of studies with range sheep in southwestern Utah.

______. 1968. Nutritive Value of Seasonal Ranges. Utah Agric. Expt. Sta. Bul. 472. 55 p.

The adequacy of different types of seasonal ranges in Utah in providing nutrient requirements, and implications for forage management and supplementation.

Cunha, T.J., R.L. Shirley, H.L. Chapman, Jr., C.B. Ammerman, et al. 1964. Minerals for Beef Cattle in Florida. Fla. Agric. Expt. Sta. Bul. 683. 60 p.

Summarization of mineral requirements, supplemental needs, and methods of supplying supplemental minerals to beef cattle in Florida.

Davenport, John W., James E. Bowns, and John P. Workman. 1973. Assessment of Sheep Losses to Coyotes--A Problem to Utah Sheepmen--A Concern of Utah Researchers. Utah Agric. Expt. Sta. Res. Rep. 7. 17 p.

Results of sheep losses--their determination and dollar value--from coyotes in southern Utah.

Davis, G.A., and R.O. Wheeler. 1970. Fall Calving in Montana. Mon. Agric. Expt. Sta. Bul. 649. 18 p.

A study of the feasibility of fall calving in Montana; general guidelines for fall versus spring calving.

Dubbs, Arthur L. 1966. Yield, Crude Protein, and Palatability of Dryland Grasses in Central Montana. Mon. Agric. Expt. Sta. Bul. 604. 18 p.

An evaluation of 16 grasses at various stages during the growing season.

Duncan, Don A., and E.A. Epps, Jr. 1958. Minor Mineral Elements and Other Nutrients on Forest Ranges in Central Louisiana. La. Agric. Expt. Sta. Bul. 516. 19 p.

Study of the adequacy of minor minerals in selected range forages and the need for trace mineral supplementation.

Duvall, V.L., and L.B. Whitaker. 1963. Supplemental Feeding Increases Beef Production on Bluestem-longleaf Pine Ranges. La. Agric. Expt. Sta. Bul. 564. 18 p.

Results of supplemental feeding studies with range cattle in central Louisiana.

Ensminger, M.E. 1970 (Rev. Ed.). The Stockman's Handbook. Interstate Print. and Pub., Danville, Ill. 957 p.

Manual of popular facts, principles, and recommendations for livestock production including range cattle and sheep.

_____. 1972. Breeding and Raising Horses. USDA Agric. Handbook 394. 79 p.

Selection, breeding, feeding, care, and management of light horses.

Fowler, Stewart H. 1969. Beef Production in the South. Interstate Print. and Pub., Danville, Ill. 858 p.

Textbook on beef cattle production with regional emphasis on the South and Southwest.

Fudge, J.F., and G.S. Fraps. 1945. The Chemical Composition of Grasses of Northwest Texas as Related to Soils and to Requirements for Range Cattle. Texas Agric. Expt. Sta. Bul. 699. 56 p.

Considers growth stage and soil effects on chemical composition of the major range grasses of the prairies and plains of northwestern Texas.

Gee, C. Kerry, Richard S. Magleby, Warren R. Bailey, Russell L. Gum, and Louise M. Arthur. 1977. Sheep and Lamb Losses to Predators and Other Causes in the Western United States. USDA Agric. Econ. Rep. 369. 41 p.

Results of special sample surveys of farmers and ranchers in western U.S. to help determine sheep and lamb losses from predation and other causes.

Gorman, John A. 1967 (5th Ed.). The Western Horse--Its Types and Training. Interstate Print. and Pub., Danville, Ill. 452 p.

Breeds and types of horses in the western states and methods of training.

Gray, J.A. 1959. Texas Angora Goat Production. Texas Agric. Ext. Bul. 926. 16 p.

Recommendations on the adaptation, selection, operation, and management of Angora goat flocks under Texas conditions.

Grunes, D.L., and H.F. Mayland. 1975. Controlling Grass Tetany. USDA Leaflet 561. 4 p.

Recognition, occurrence, prevention, and treatment of grass tetany in cattle.

Guilbert, H.R., and G.H. Hart. 1952. California Beef Production. Calif. Agric. Expt. Sta. Manual 2. Variously paged.

A manual on all phases of beef production under California conditions.

Halls, L.K., O.M. Hale, and F.E. Knox. 1957. Seasonal Variation in Grazing Use, Nutritive Content, and Digestibility of Wiregrass Forage. Ga. Agric. Expt. Sta. Tech. Bul. N.S. 11. 28 p.

Studies on the nutritional adequacy of wiregrass forage in Georgia.

Halls, L.K., R.H. Hughes, R.S. Rummell, and B.L. Southwell. 1964. Forage and Cattle Management in Longleaf-slash Pine Forests. USDA Farm. Bul. 2199. 25 p.

Recommendations on forage and cattle management on longleaf-slash pine forests of the South.

Hamilton, John W., and Carl S. Gilbert. 1972. Composition of Wyoming Range Plants and Soils. Wyo. Agric. Expt. Sta. Res. J. 55. 14 p.

Levels of important mineral elements in selected forage plants.

Harris, Lorin E. 1968. Range Nutrition in an Arid Region. Utah State Univ. Honor Lecture 36. 95 p.

An honor lecture emphasizing the development of range nutrition research methods and the results of range nutrition experimentation on arid Intermountain ranges.

Harris, Lorin E., C. Wayne Cook, and L.A. Stoddart. 1956. Feeding Phosphorus, Protein, and Energy Supplements to Ewes on Winter Ranges of Utah. Utah Agric. Expt. Sta. Bul. 398. 28 p.

The effectiveness and economy of supplementing ewes winter grazing salt-desert ranges of western Utah.

Hatch, C.F., A.B. Nelson, R.D. Pieper, and R.P. Kromann. 1968. Chemical Composition of Grasses at the Fort Stanton Experimental Range in South-central New Mexico. N. Mex. Agric. Expt. Sta. Res. Rep. 148. 11 p.

Seasonal changes in chemical composition and dry matter digestibility of grasses at the Fort Stanton station.

Hickman, O. Eugene. 1975. Seasonal Trends in the Nutritive Content of Important Range Forage Species Near Silver Lake, Oregon. USDA, For. Serv. Res. Paper PNW-187. 32 p.

Seasonal nutrient trends of several prominent grasses, forbs, and shrubs as affected by vegetation types and season of year.

Huston, J.E., Maurice Shelton, and W.C. Ellis. 1971. Nutritional Requirements of the Angora Goat. Texas Agric. Expt. Sta. Bul. 1105.

Considers nutrient requirements of Angora goats, their diets and likely nutrient deficiencies, and suggested means for supplying proper nutrition.

Maddox, L.A., Jr. 1970. Keys to Profitable Cow-calf Operations. Texas Agric. Ext. Serv. Misc. Pub. 956. 20 p.

Management and operational practices for economical, efficient cow-calf production.

_____. 1971. (Coord.) Beef Cattle Management During Drouth. Texas Agric. Ext. Bul. 1108. 32 p.

Economic consequences, financial strategy, cow-calf management, feeding suggestions, and range and pasture management under severe drought conditions.

Marsh, H., K.F. Swingle, R.R. Woodward, G.F. Payne, et al. 1959. Nutrition of Cattle on an Eastern Montana Range as Related to Weather, Soil, and Forage. Mon. Agric. Expt. Sta. Bul. 549. 91 p.

Results of long-time studies at the U.S. Range Livestock Experiment Station with particular reference to different grazing intensities.

Melton, A.A., J.H. Jones, and J.K. Riggs. 1961. Frequency of Supplemental Feeding for Range Cattle, Trans-Pecos Area. Texas Agric. Expt. Sta. Prog. Rep. 2188. 4 p.

Results of different frequencies of cottonseed cake supplementation with range cattle in western Texas.

Murray, R.B., H.F. Mayland, and P.J. Van Soest. 1978. Growth and Nutritional Value to Cattle of Grasses on Cheatgrass Range in Southern Idaho. USDA, For. Serv. Res. Paper INT-199. 57 p.

Trends in nutrient contents and digestibility fractions for seven grass species and causes of these trends; predicting probable mineral deficiencies.

Natl. Res. Council, Comm. on Natural Resources. 1976. Urea and Other Nonprotein Nitrogen Compounds in Animal Nutrition. Natl. Acad. Sci., Washington, D.C. 120 p.

A review of the utilization of supplemental nonprotein nitrogen compounds by ruminant and nonruminant animals.

Natl. Res. Council, Subcomm. on Feed Composition, and Canada Dept. Agric., Comm. on Feed Composition. 1971. Atlas of Nutritional Data on United States and Canadian Feeds. Natl. Acad. Sci., Washington, D.C. 772 p.

Tables of feed composition, alphabetically arranged, for all feeds used in North America.

Nelson, A.B., C.H. Herbel, and H.M. Jackson. 1970. Chemical Composition of Forage Species Grazed by Cattle on an Arid New Mexico Range. N. Mex. Agric. Expt. Sta. Bul. 561. 33 p.

Results of studies at the Jornada Experimental Range; chemical composition of grasses, forbs and shrubs in relation to animal nutrient requirements.

Nelson, A.B., L.S. Pope, Robert MacVicar, A.E. Darlow, and D.F. Stephens. 1954. Self-Feeding Salt and Cottonseed Meal to Beef Cattle. Okla. Agric. Expt. Sta. Bul. 440. 14 p.

Using salt to regulate meal intake by beef cattle.

Nelson, A.B., W.D. Gallup, O.B. Ross, and A.E. Darlow. 1955. Supplemental Phosphorus Requirement of Range Beef Cattle in North Central and Southeastern Oklahoma. Okla. Agric. Expt. Sta. Tech. Bul. 54. 29 p.

Phosphorus trends in range grasses; growth and reproductive performance of cattle on different levels of phosphorus intake; minimum phosphorus requirements.

Orcutt, E.P. 1956. Wintering Montana Ewes. Mon. Agric. Ext. Bul. 291. 31 p.

Recommendations for winter feeding and management of ewes in Montana.

Parker, E.E., J.D. Wallace, A.B. Nelson, and R.D. Pieper. 1974. Effects of Cottonseed Meal Supplement and Age at First Calving on Performance of Range Cattle. N. Mex. Agric. Expt. Sta. Bul. 627. 24 p.

Summarization of several winter supplemental feeding trials with weanling calves and producing cows at the Fort Stanton Experimental Ranch.

Parker, Karl G. 1961. Range Practices that Help Prevent Water-belly. Mon. Agric. Ext. Circ. 278. 8 p.

Occurrence, vegetation and mineral relationships, and prevention of urolithiasis in steers.

Pieper, Rex D. 1970. Species Utilization and Botanical Composition of Cattle Diets on Pinyon-juniper Grassland. N. Mex. Agric. Expt. Sta. Bul. 566. 16 p.

Results of studies at the Fort Stanton Range Research Station in New Mexico.

Pope, L.S., D. Stephens, R.D. Humphrey, Robert MacVicar, and O.B. Ross. 1956. Wintering and Fattening Steers on Native Grass. Okla. Agric. Expt. Sta. Bul. 474. 31 p.

Effects of different supplements, methods of feeding, and type of range on animal performance.

Reynolds, E.B., J.M. Jones, J.H. Jones, J.F. Fudge, and R.J. Kleberg, Jr. 1953. Methods of Supplying Phosphorus to Range Cattle in South Texas. Texas Agric. Expt. Sta. Bul. 773. 16 p.

Results of studies on phosphating rangelands and on supplying supplemental phosphorus by adding to drinking water, feeding in self-feeders, and grazing phosphated pasture forage.

Rice, F.J., R.R. Woodward, J.R. Quesenberry, and F.S. Willson. 1961. Fertility of Beef Cattle Raised Under Range Conditions. Mon. Agric. Expt. Sta. Bul. 461. 8 p.

Study of cow fertility records at the U.S. Range Livestock Experiment Station and factors affecting size of calf crop.

Robertson, Joseph H., and Clark Torell. 1958. Phenology as Related to Chemical Composition of Plants and to Cattle Gains on Summer Ranges in Nevada. Nev. Agric. Expt. Sta. Bul. (Tech.) 197. 38 p.

Investigation of factors influencing decline in chemical composition of plants that cause decline in gains or weight losses in late summer.

Savage, D.A., and V.G. Heller. 1947. Nutritional Qualities of Range Forage Plants in Relation to Grazing With Beef Cattle on the Southern Plains Experimental Range. USDA Tech. Bul. 943. 61 p.

Chemical composition of native and introduced grasses; and effects of nutrient levels, type of forage plants, and supplementation on cattle gains.

Skovlin, Jon M. 1967. Fluctuations in Forage Quality on Summer Range in The Blue Mountains. USDA, For. Serv. Res. Paper PNW-44. 20 p.

Forage quality decline interpreted in light of plant development, dates of phenological events, curing conditions, and secondary fall regrowth.

Sneva, Forrest A., L.R. Rittenhouse, and V.E. Hunter. 1977. Stockwater's Effect on Cattle Performance on the High Desert. Ore. Agric. Expt. Sta. Bul. 625. 7 p.

Effects of frequency of watering and trailing distance to water on cow, yearling, and calf performance.

Spitzer, J.C., J.N. Wiltbank, and D.G. LeFever. 1975. Increase Beef Cow Productivity by Increasing Reproductive Performance. Colo. Agric. Expt. Sta. Gen. Ser. 949. 9 p.

Practical measures and benefits of increasing reproductive performance of beef cattle.

Spurlock, G.M., W.C. Weir, G.E. Bradford, and Reuben Albaugh. 1969. Production Practices for California Sheep. Calif. Agric. Expt. Sta. Manual 40. 92 p.

A manual presenting the principal aspects of sheep production in California.

Streeter, C.L., D.C. Clanton, and O.E. Hoehne. 1968. Influence of Advance in Season on Nutritive Value of Forage Consumed by Cattle Grazing Western Nebraska Native Range. Neb. Agric. Expt. Sta. Res. Bul. 227. 21 p.

Results of a study on factors affecting dietary intake of grazing cattle and their nutritional implications.

USDA, Agric. Res. Serv. 1965. The Effect of Soils and Fertilizers on the Nutritional Quality of Plants. USDA Agric. Info. Bul. 299. 24 p.

The status of our knowledge and some continuing problems.

USDA, Inter-Agency Work Group on Range Production. 1974. Opportunities to Increase Red Meat Production From Ranges of the United States, Phase I--Non-Research. USDA, Washington, D.C. 100 p.

Current and prospective demands for range for livestock grazing, capability of rangelands to provide projected demands, and management opportunities for increasing range forage supply.

Van Horn, J.L., G.F. Payne, F.S. Willson, J. Drummond, O.O. Thomas, and F.A. Branson. 1959. Protein Supplementation of Range Sheep. Mon. Agric. Expt. Sta. Bul. 547. 15 p.

Effect of various protein levels of supplemental feeding on winter range upon lamb and wool production.

Wagnon, Kenneth A. 1965. Social Dominance in Range Cows and Its Effect on Supplemental Feeding. Calif. Agric. Expt. Sta. Bul. 819. 32 p.

Behavioral traits of cattle at different ages at the San Joaquin Experimental Range and implications in supplemental feeding practices.

Wagnon, K[enneth].A., and F.D. Carroll. 1966. Reproduction Difficulties in Range Beef Cattle. Calif. Agric. Expt. Sta. Bul. 822. 28 p.

Magnitude and causes of reproductive losses in experimental range beef cattle herds at the San Joaquin Experimental Range.

Watkins, W.E., and W.W. Repp. 1964. Influence of Location and Season on the Composition of New Mexico Range Grasses. N. Mex. Agric. Expt. Sta. Bul. 486. 33 p.

Seasonal variations in chemical composition of 10 range grasses and losses of nutrients from weathering, leaching, and other causes.

Watson, Ivan. 1958. Range Sheep Production. N. Mex. Agric. Ext. Cir. 290. 23 p.

Recommendations for range sheep management in New Mexico.

Waymack, Lester B., and Albert M. Lane. 1975. Feeding the Arizona Horse. Ariz. Agric. Ext. Bul. A-84. 13 p.

A nutrition guide for horse owners in Arizona and the Southwest.

Weir, W.C., and D.T. Torell. 1966. Supplemental Feeding of Sheep Grazing on Dry Range. Calif. Agric. Expt. Sta. Bul. 832.

Amount and kinds of feeds and frequency of feeding for proper supplementation of weaning lambs grazing dry annual-range forage.

Weisenburger, R.D. 1976. Beef Cow-calf Manual. Alberta Agric. Agdex 420/10. Variously paged.

A comprehensive, semitechnical manual on practical cow-calf production in Alberta and related areas.

Willson, F.S., O.O. Thomas, and N.A. Jacobsen. 1972. Beefing Up Your Cattle Profits Through Nutrition and Management. Mon. Agric. Ext. Bul. 339. 15 p.

Popular recommendations on the nutrition and management of beef cattle breeding herds in Montana.

Range Wildlife[49]

Advisory Comm. On Predation Control (Stanley A. Cain, Chm.). 1972. Predator Control--1971. Inst. for Environ. Quality, Univ. of Mich., Ann Arbor, Mich. 207 p.

The controversial "Cain Report," a review and analysis of predator control and associated animal control programs and policies in the U.S.

Anderson, Chester C. 1958. The Elk of Jackson Hole. Wyo. Game and Fish Comm. Bul. 10. 184 p.

Life history, ecology, and management of the elk of Jackson Hole.

Ariz. Game and Fish Dept. 1977. The Arizona Whitetail Deer. Ariz. Game and Fish Dept. Spec. Rep. 6. 108 p.

Biology, life history, distribution, habitat, and management of whitetail deer in Arizona.

Bear, George D., and George W. Jones. 1973. History and Distribution of Bighorn Sheep in Colorado. Colo. Div. Wildlife, Denver. 232 p.

Status and management of bighorn sheep in Colorado.

Boggess, Edward K., and F. Robert Henderson. 1977. Kansas Big Game and Its Management. Kan. Agric. Ext. Cir. 566. 16 p.

Life history, distribution, and management of big game in Kansas.

Brandborg, Stewart M. 1955. Life History and Management of the Mountain Goat in Idaho. Idaho Dept. Fish and Game Wildl. Bul. 2. 142 p.

Status of the mountain goat in Idaho to 1955.

Brown, B.R. 1961. The Black-tailed Deer of Western Washington. Wash. State Game Dept. Biol. Bul. 13. 124 p.

Life history, ecology, and management of black-tailed deer in western Washington.

Buechner, Helmut K. 1960. The Bighorn Sheep in the United States, Its Past, Present, and Future. Wildl. Monogr. 4. 174 p.

History, biology, distribution, habitat requirements, and management of bighorn sheep in the U.S.

Casebeer, R.L., M.J. Rognrud, and S.M. Brandborg. 1950. Rocky Mountain Goats in Montana. Mon. Fish and Game Comm. Wildl. Rest. Div. Bul. 5. 107 p.

Biology, life history, and management of Rocky Mountain goats in Montana.

Compton, Thomas. 1975. Mule Deer-elk Relationships in the Western Sierra Madre Area of Southcentral Wyoming. Wyo. Game and Fish Dept. Wildl. Tech. Rep. 1. 125 p.

Study of comparative distribution, food habits, and management of deer and elk in Wyoming.

Craighead, John J., Gerry Atwell, and Bart W. O'Gara. 1972. Elk Migrations In and Near Yellowstone National Park. Wildl. Monogr. 29. 48 p.

Study to locate and delimit the summer and winter ranges and migrations between these ranges.

Evans, Keith E., and George E. Probasco. 1977. Wildlife of the Prairies and Plains. USDA, For. Serv. Gen. Tech. Rep. NC-29. 18 p.

Summary of wildlife resources and wildlife habitat management in the principal grassland types of the U.S.

Fierro G., Luis Carlos. 1977. Influence of Livestock Grazing on the Regrowth of Crested Wheatgrass for Winter Use by Mule Deer. Utah Div. Wildl. Resources Pub. 77-17. 66 p.

Seasonal use of crested wheatgrass by mule deer; effect of livestock grazing during the growing season on regrowth for winter use by mule deer.

Flinders, Jerran T., and Richard M. Hansen. 1972. Diets and Habitats of Jackrabbits in Northeastern Colorado. Colo. State Univ. Range Science Dept. Sci. Ser. 12. 29 p.

Comparison of the yearly and seasonal diets and feeding habits of the black-tailed and white-tailed jackrabbits in a shortgrass ecosystem.

Garrett, J.R., G.J. Pon, and D.J. Arosteguy. 1970. Economics of Big Game Resource Use in Nevada. Nev. Agric. Expt. Sta. Bul. 25. 22 p.

Demand estimation, resource valuation, and evaluation of rehabilitation projects for big game in Nevada.

Giles, Robert H., Jr. 1978. Wildlife Management. W.H. Freeman and Co., San Francisco. 416 p.

A new, comprehensive textbook on wildlife management.

Goodwin, Gregory A. 1975. Seasonal Food Habits of Mule Deer in Southeastern Wyoming. USDA, For. Serv. Res. Note RM-287. 4 p.

Evaluation of seasonal diets, available foods, and vegetation types.

Hahn, H.C. 1945. The White-tailed Deer in the Edwards Plateau Region of Texas. Texas Game, Fish, and Oyster Comm., Austin. 50 p.

Ecology, life history, distribution, habitat requirements, and management.

Halls, Lowell K. (Coord.). 1969. White-tailed Deer in the Southern Forest

Habitat, Proceedings of a Symposium at Nacogdoches, Texas, March 25-26, 1969. USDA, Southern For. Expt. Sta. 130 p.

A symposium to review known information and express new ideas on the background, characteristics, and management of whitetail deer and their habitat in the South.

Hansen, R.M., and J.T. Flinders. 1969. Food Habits of North American Hares. Colo. State Univ. Range Sci. Dept. Sci. Ser. 1. 18 p.

Review of literature pertaining to the food habits of North American hares.

Harlow, R.F., and F.K. Jones. 1965. The White-tailed Deer in Florida. Fla. Game and Fresh Water Fish Comm. Tech. Bul. 9. 240 p.

Life history, ecology, and management.

Harper, James A., Joseph H. Harn, Wallace W. Bentley, and Charles F. Yocom. 1967. The Status and Ecology of the Roosevelt Elk in California. Wildl. Monogr. 16. 49 p.

Physical characteristics, food habits, biology, and social behavior of the Roosevelt elk.

Hickey, William O. 1976. Bighorn Sheep Ecology. Idaho Dept. Fish and Game, Boise. 54 p.

Habitat needs, distribution, life history, and management of bighorn sheep in Idaho.

Hoover, R.L., C.E. Till, and S. Ogilvie. 1959. The Antelope of Colorado. Colo. Dept. Game and Fish Tech. Bul. 4. 110 p.

Life history, ecology, and management of antelope in Colorado.

Howard, Volney W., Jr., Charles T. Engelking, E. Dwain Glidwell, and John E. Wood. 1973. Factors Restricting Pronghorn Increase on the Jornada Experimental Range. N. Mex. Agric. Expt. Sta. Res. Rep. 245. 13 p.

A study of pronghorn habitat requirements and their management implications.

John, Rodney T. 1968. Utah's Desert Bighorn. Utah State Div. Fish and Game Pub. 68-9. 19 p.

Biology, life history, distribution, ecology, and management.

Kramer, A. 1972. A Review of the Ecological Relationships Between Mule and White-tailed Deer. Alberta Dept. Lands and Forests, Fish and Wildl. Div. Occas. Paper 3. 54 p.

Comparison of habitat use, food habits, and ecology of mule and white-tailed deer.

Kufeld, Roland C., O.C. Wallmo, and Charles Feddema. 1973. Foods of the Rocky Mountain Mule Deer. USDA, For. Serv. Res. Paper RM-111. 31 p.

A review of literature on food habits of the Rocky Mountain deer and a classification of the relative importance of deer forage plant species.

Lyon, L. Jack. 1966. Problems of Habitat Management for Deer and Elk in Northern Forests. USDA, For. Serv. Res. Paper INT-24. 15 p.

An evaluation of habitat problems and research needs.

Mackie, Richard J. 1970. Range Ecology and Relations of Mule Deer, Elk and Cattle in the Missouri River Breaks, Montana. Wildl. Monogr. 20. 79 p.

Results of a study of mule deer-elk-cattle relationships with emphasis on competitive range use and food habits.

Merrill, Leo B., Gerald W. Thomas, W.T. Hardy, E.B. Keng, et al. 1957. Livestock and Deer Ratios for Texas Range Lands. Texas Agric. Expt. Sta. Misc. Pub. 221. 9 p.

A review of pertinent research and field observations; formulation of animal unit equivalence for cattle, sheep, goats, and whitetail deer.

Midwest Fish and Wildl. Conf. 1970. White-tailed Deer in the Midwest. A Symposium. USDA, For. Serv. Res. Paper NC-39. 34 p.

Papers on the status of knowledge and research needs on white-tailed deer in the Midwest.

Moser, Clifford A. 1962. The Bighorn Sheep of Colorado. Colo. Dept. Game and Fish Tech. Pub. 10. 49 p.

Life history, ecology, and management of bighorn sheep in Colorado.

Mussehl, Thomas W., and F.W. Howell (Eds.). 1971. Game Management in Montana. Mon. Fish and Game Comm., Helena. 238 p.

Individually authored chapters on the life history, ecology, and management of big game, game bird, and fur animals in Montana.

Natl. Res. Council, Comm. on Agric. Land Use and Wildl. Resources. 1971. Land Use and Wildlife Resources. Natl. Acad. Sci., Washington, D.C. 262 p.

Guidelines for the consideration of wildlife resources in land use planning and management.

Natl. Res. Council, Subcomm. on Vetebrate Pests. 1970. Principles of Plant and Animal Pest Control. Volume 5. Vertebrate Pests: Problems and Control. Natl. Acad. Sci., Washington, D.C. 153 p.

Basic problems, principles, and research needs for controlling vertebrate pests.

Nelson, Lewis, Jr., and Jon K. Hooper. 1975. California Big Game and Its Management. Calif. Agric. Ext. Leaflet 2223. 24 p.

Life history, distribution, and management of big game in California.

N. Mex. Dept. Game and Fish. 1967. New Mexico Wildlife Management. N. Mex. Dept. Game and Fish, Santa Fe. 250 p.

Life history, ecology, and management of the principal wildlife species of New Mexico.

Page, Leslie Andrew (Ed.). 1976. Wildlife Diseases. Plenum Pub. Corp., New York. 691 p.

A technical encyclopedia of wildlife diseases.

Park, E. 1969. The World of the Bison. J.B. Lippincott, Philadelphia. 161 p.

Popular treatment of the history, physical characteristics, behavior, and habitat of the bison in North America.

Phillips, W.E. 1976. The Conservation of the California Tule Elk. Univ. Alberta Press, Edmonton. 120 p.

A socioeconomic study of the endangered status and public policies needed to increase the population of the California tule elk.

Ramsey, Charles. c1970. Texotics. Texas Parks and Wildl. Dept. Bul. 49. 46 p.

The introduction, adaptation, and management of introduced game species in Texas.

______. 1970. Wild Game of Texas. Texas Agric. Ext. Bul. 150. 15 p.

Popular treatment of life histories, distribution, and management of wild game species in Texas.

Reynolds, Temple A., Jr. 1960. The Mule Deer--Its History, Life History and Management in Utah. Utah State Dept. Fish and Game Dept. Inf. Bul. 60-4. 32 p.

A popular publication pertaining to the mule deer in Utah.

Richardson, Arthur H. 1971. The Rocky Mountain Goat in the Black Hills. S. Dak. Dept. Game, Fish, and Parks Bul. 2. 24 p.

Life history, distribution, habitat, and management of the Rocky Mountain goat in the Black Hills.

Richardson, Arthur H., and Lyle E. Petersen. 1974. History and Management of South Dakota Deer. S. Dak. Dept. Game, Fish and Parks Bul. 5. 113 p.

A popular to semitechnical publication on the management of mule and whitetail deer resources in South Dakota.

Rue, Leonard Lee, III. 1962. The World of the White-tailed Deer. J.B. Lippincott Co., Philadelphia. 134 p.

Popular publication on the physical characteristics, behavior, habitat, and management of the white-tailed deer.

Russell, T.P. 1964. Antelope of New Mexico. N. Mex. Dept. Game and Fish. Bul. 12. 103 p.

Life history, distribution, food habits, and management of antelope in New Mexico.

Russo, John P. 1956. The Desert Bighorn Sheep in Arizona. Ariz. Game and Fish Dept. Wildl. Bul. 1. 153 p.

Life history and management problems.

______. 1964. The Kaibab North Deer Herd--Its History, Problems and Management. Ariz. Game and Fish Dept. Wildl. Bul. 7. 195 p.

History and population dynamics, habitat use, and habitat improvements related to the Kaibab North Deer Herd to 1962.

Schmidt, John L., and Douglas L. Gilbert. 1978. Big Game of North America, Ecology and Management. Stackpole Books, Harrisburg, Pa. 494 p.

A new, comprehensive reference manual on big game ecology and management in North America.

Schneeweis, James C., Keith E. Severson, Lyle E. Peterson, Theron E. Schenck III, et al. 1972. Food Habits of Deer in the Black Hills. S. Dak. Agric. Expt. Sta. Bul. 606. 35 p.

Determination of the principal plants used by mule and white-tailed deer in fall, winter, and spring.

Severson, Keith, Morton May, and William Hepworth. 1968. Food Preferences, Carrying Capacities, and Forage Competition Between Antelope and Domestic Sheep in Wyoming's Red Desert. Wyo. Agric. Expt. Sta. Sci. Monog. 10. 51 p.

Analyses of food habits of antelope and sheep confined in pastures when grazed together and separately.

Short, Henry L., and Clay Y. McCulloch. 1977. Managing Pinyon-juniper Ranges for Wildlife. USDA, For. Serv. Gen. Tech. Rep. RM-47. 10 p.

A review of pinyon-juniper woodlands and associated animal communities in the western United States and management advantageous to wildlife.

Short, Henry L., Robert M. Blair, and E.A. Epps, Jr. 1975. Composition and Digestibility of Deer Browse in Southern Forests. USDA, For. Serv. Res. Paper SO-111. 10 p.

Emphasis given to seasonal changes in leaves and twigs of important deer browse species.

Smith, Arthur D., and Dean D. Doell. 1968. Guides to Allocating Forage Between Cattle and Big Game Winter Range. Utah State Div. Fish and Game Pub. 68-11. 32 p.

Determination of season and intensity of cattle grazing that would minimize woody plant browsing but allow full use of herbaceous species.

Smith, Dixie R. 1961. Competition Between Cattle and Game on Elk Winter Range. Wyo. Agric. Expt. Sta. Bul. 377. 16 p.

With implications for management in Wyoming.

Smith, Dwight R. 1954. The Bighorn Sheep in Idaho: Its Status, Life History, and Management. Idaho Dept. Fish and Game Wild. Bul. 1. 154 p.

Semitechnical report on Bighorn sheep in Idaho.

Spillett, J. Juan, Jessop B. Low, and David Sill. 1967. Livestock Fences--How They Influence Pronghorn Antelope Movements. Utah Agric. Expt. Sta. Bul. 470. 79 p.

A study of fence types and fence devices to permit antelope crossing while satisfactorily holding livestock.

Stelfox, John G. 1976. Range Ecology of Rocky Mountain Bighorn Sheep. Can. Wildl. Serv. Rep. Ser. 39. 50 p.

Study of habitat requirements and use by Rocky Mountain bighorn sheep.

Sundstrom, Charles, William P. Hepworth, and Kenneth L. Diem. 1973. Abundance, Distribution, and Food Habits of the Pronghorn. Wyo. Game and Fish Comm. Bul. 12. 61 p.

With implications for antelope management in Wyoming.

Swank, W.G. 1958. The Mule Deer in Arizona Chaparral and an Analysis of Other Important Deer Herds. Ariz. Game and Fish Dept. Wildl. Bul. 3. 109 p.

Deer foods, biology, and population dynamics of selected deer herds.

Taber, Richard D., and Raymond F. Dasmann. 1958. The Black-Tailed Deer of the Chaparral. Calif. Dept. Fish and Game Bul. 8. 163 p.

Its life history and management in the northern Coast Range of California.

Teague, Richard D. (Ed.). 1971. A Manual of Wildlife Conservation. The Wildlife Society, Washington, D.C. 206 p.

A popular review of wildlife management and conservation principles.

Teer, James G. 1965 (Rev.). Texas Deer Herd Management Problems and Principles. Texas Parks and Wildl. Dept. Bul. 44. 69 p.

Popular bulletin on the problems and biological principles of sound deer herd management in Texas.

Teer, James G., Jack W. Thomas, and Eugene A. Walker. 1965. Ecology and Management of White-tailed Deer in the Llano Basin of Texas. Wildl. Monogr. 15. 62 p.

Study of population trends, structure, and welfare of deer; ecological and management factors influencing the deer herd and its management.

Thorne, Tom, and Gary Butler. 1976. Comparison of Pelleted, Cubed and Baled Alfalfa Hay as Winter Feed for Elk. Wyo. Game and Fish Dept. Wildl. Tech. Rep. 6. 38 p.

Results of a feeding study in overwintering elk.

Trefethen, James B. (Ed.). 1975. The Wild Sheep in Modern North America. The Winchester Press, New York. 302 p.

Life history, distribution, habitat requirements, and management of wild sheep in North America.

Tueller, Paul T., and Leslie A. Monroe. 1975. Management Guidelines for Selected Deer Habitats in Nevada. Nev. Agric. Expt. Sta. Rep. 104. 185 p.

Recommendations for improvement and management of different deer habitats in Nevada.

Turner, G.T., R.M. Hansen, V.H. Reid, H.P. Tietjen, and A.L. Ward. 1973. Pocket Gophers and Colorado Mountain Rangeland. Colo. Agric. Expt. Sta. Bul. 554. 90 p.

Biology of the pocket gopher, its influence on range vegetation and soils, and effective and practical control methods.

Van Wormer, Joe. 1969. The World of the American Elk. J.B. Lippincott Co., Philadelphia. 159 p.

Life history, physical characteristics, and ecology of the American elk.

______. 1969. The World of the Pronghorn. J.B. Lippincott Co., Philadelphia. 191 p.

Life history, physical characteristics, ecology, and management of the pronghorn antelope.

Workman, Gar W., and Jessop B. Low (Eds.). 1976. Mule Deer Decline in the West: A Symposium. Utah State University, Logan, Utah, April, 1976. 134 p.

Papers in a symposium presenting current knowledge pertaining to mule deer populations, research, and management.

Range Watersheds

Amer. Soc. Civil Eng. 1975. Watershed Management. Amer. Soc. Civil Eng., New York. 781 p.

A comprehensive textbook on hydrology as related to watershed management, with range watershed implications.

Anderson, Henry W., Marvin D. Hoover, and Kenneth G. Reinhart. 1976. Forests and Water: Effects of Forest Management on Floods, Sedimentation, and Water Supply. USDA, For. Serv. Gen. Tech. Rep. PSW-18. 115 p.

A comprehensive review of literature; includes nearly 600 references.

Brown, Darrell. 1977. Handbook: Equipment for Reclaiming Strip Mined Land. USDA, For. Serv., Equip. Dev. Center, Missoula, Mon. 58 p.

Summarization of essential facts about equipment used for coal surface mining and reclamation, including their description, source, and functions.

Brown, Harry E., Malchus B. Baker, Jr., James J. Rogers, Warren P. Clary, et al. 1974. Opportunities for Increasing Water Yields and Other Multiple Use Values on Ponderosa Pine Forest Lands. USDA, For. Serv. Res. Paper RM-129. 36 p.

With special emphasis on the results of experimental land treatments on the Beaver Creek Pilot Watershed near Flagstaff.

Colo. State Univ., Environ. Resources Center. 1975. Revegetation of High-Altitude Disturbed Lands: Proceedings of a Workshop at Fort Collins, January 31 to February 1, 1974. Environ. Resources Center, Colo. State Univ., Fort Collins.

Status of our knowledge and unanswered problems on revegetating high-altitude, disturbed watersheds.

Cook, C. Wayne, Robert M. Hyde, and Phillip L. Sims. 1974. Revegetation Guidelines for Surface Mined Areas. Colo. State Univ. Range Sci. Ser. 16. 70 p.

Summary of studies concerned with revegetation methods and techniques involving the use of mulches, fertilizers, herbicides, adapted species, and planting methods.

Doty, Robert D. 1971. Contour Trenching Effects on Streamflow from a Utah Watershed. USDA, For. Serv. Res. Paper INT-95. 19 p.

Results of contour trenching on annual streamflow, characteristics of spring streamflow, and low streamflow on two Utah watersheds.

Ffolliott, Peter F., and David B. Thorud. 1974. Vegetation Management for Increased Water Yield in Arizona. Ariz. Agric. Expt. Sta. Tech. Bul. 215. 38 p.

Organization, analysis, and summarization of knowledge of hydrologic processes and principles applicable to Arizona watershed management and vegetation resources.

Frank, Ernest C., Harry E. Brown, and J.R. Thompson. 1975. Hydrology of Black Mesa Watersheds, Western Colorado. USDA, For. Serv. Gen. Tech. Rep. RM-13. 11 p.

A study of sediment-ground cover relationships.

Gary, Howard L. 1975. Watershed Management Problems and Opportunities for the Colorado Front Range Ponderosa Pine Zone: The Status of Our Knowledge. USDA, For. Serv. Res. Paper RM-139. 32 p.

Includes guidelines for maintaining satisfactory watershed conditions and practical alternatives for increasing water supplies.

Heady, Harold F., Donna H. Falkenborg, and J. Paul Riley (Ed.). 1976. Watershed Management on Range and Forest Lands: Proceedings of the Fifth Workshop of the United States/Australia Rangelands Panel, Boise, Idaho, June 15-22, 1975. Utah Water Res. Lab., Utah State Univ., Logan. 222 p.

Papers on hydrology and applied watershed management.

Heede, Burchard H. 1976. Gully Development and Control: The Status of Our Knowledge. USDA, For. Serv. Res. Paper RM-169. 42 p.

Mechanics, processes, and morphology of gully formation and the design of gully control structures and land treatments.

Horton, Jerome S., and C.J. Campbell. 1974. Management of Phreatophyte and Riparian Vegetation for Maximum Multiple Use Values. USDA, For. Serv. Res. Paper RM-117. 23 p.

Status of knowledge about environmental relationships of vegetation along southwestern water courses and the impacts of vegetation management to reduce evapotranspiration.

Hutnik, Russell J., and Grant Davis. 1973. Ecology and Reclamation of Devasted Land, 2 Vols. Gordon and Breach, New York. 538 and 504 p.

Revegetation of strip-mined areas, with emphasis on subhumid areas.

Leaf, Charles F. 1975. Watershed Management in the Central and Southern Rocky Mountains: A Summary of Our Knowledge by Vegetation Types. USDA, For. Serv. Res. Paper RM-142. 28 p.

Summarization of knowledge of hydrology of these lands, hydrologic principles important for multiresource management, and additional information needed.

_____. 1975. Watershed Management in the Rocky Mountain Subalpine Zone: The Status of Our Knowledge. USDA. For. Serv. Res. Paper RM-137. 31 p.

Covers forest hydrology, watershed management practices, treatment simulation modeling, research needs, and land use planning implications.

Martinelli, M., Jr. 1975. Water-yield Improvement From Alpine Areas: The Status of Our Knowledge. USDA, For. Serv. Res. Paper RM-138. 16 p.

Snowpack management in the alpine zone.

May, Morton, Robert Lang, Leandro Lujan, Peter Jacoby, and Wesley Thompson. 1971. Reclamation of Strip Mine Spoil Banks in Wyoming. Wyo. Agric. Expt. Sta. Res. J. 51. 32 p.

Study of plant species adaptation, fertilization, mulching, snow fencing, and mechanical soil treatments on vegetation establishment and growth.

Meyn, R.L., E.S. Sundberg, R.P. Young, and R.L. Hodder. 1976. Research on Reclamation of Surface Mined Lands at Colstrip, Montana. Mon. Agric. Expt. Sta. Res. Rep. 101. 84 p.

Summary of revegetation trials.

Orr, Howard K. 1975. Watershed Management in the Black Hills: The Status of Our Knowledge. USDA, For. Serv. Res. Paper RM-141. 12 p.

Ecology, water yields, and watershed management research and problems unique to the Black Hills.

Packer, Paul E. 1974. Rehabilitation Potentials and Limitations of Surface-mined Land in the Northern Great Plains. USDA, For. Serv. Gen. Tech. Rep. INT-14. 44 p.

Data acquisition, classification, analysis, and interpretation needed in prediction of rehabilitation site potentials.

Rauzi, Frank, C.L. Fly, and E.J. Dyksterhuis. 1968. Water Intake on Mid-continental Rangelands as Influenced by Soil and Plant Cover. USDA Tech. Bul. 1390. 58 p.

Results of water intake studies in the Northern and Central Great Plains.

Rich, Lowell R., and J.R. Thompson. 1974. Watershed Management in Arizona's Mixed Conifer Forests: The Status of Our Knowledge. USDA, For. Serv. Res. Paper RM-130. 15 p.

Summarization of water yield studies in Arizona's mixed conifer vegetation with implications to alternative forest management practices.

Sampson, George R., Harold E. Worth, and Dennis M. Donelly. 1974. Chaparral Conversion Potential in Arizona: Part 1. Water Yield Response and Effects on Other Resources. USDA, For. Serv. Res. Paper RM-126. 36 p.

Opportunities, methods, safeguards, and results of chaparral conversion in Arizona.

Satterlund, Donald R. 1972. Wildland Watershed Management. Ronald Press Co., New York. 370 p.

A textbook presenting the fundamental theory and basic practices of managing wildlands for water.

Sturges, David L. 1975. Hydrologic Relations on Undisturbed and Converted Big Sagebrush Lands: The Status of Our Knowledge. USDA, For. Serv. Res. Paper RM-140. 23 p.

A review of hydrology, management, and treatment practices on big sagebrush watersheds.

USDA, Soil Cons. Serv. 1973 (Rev.). How to Control a Gully. USDA Farm Bul. 2171. 16 p.

Practical recommendations on gully control practices and structures.

Vories, Kimery C. (Ed.). 1976. Reclamation of Western Surface Mined Lands. Ecology Consultants, Fort Collins, Colo. 152 p.

Proceedings of a symposium on reclamation of mining disturbed sites.

Wright, Robert A. (Ed.). 1978. The Reclamation of Disturbed Arid Lands. Univ. New Mex. Press, Albuquerque. 196 p.

Proceedings of a symposium held in Denver, Colo., on February 23-24, 1977; focus of papers is on current and recent research on the reclamation of lands disturbed primarily by mining activity in the arid United States.

Ranch Economics

Allen, Herbert R., and Rex D. Helfinstine. 1969. An Economic Analysis of Ranch Organization in Central South Dakota. S. Dak. Agric. Expt. Sta. Tech. Bul. 33.

Study of ranch organization, pasture improvement and management programs, and adjustments during pasture renovation to maximize net returns.

Amer. Soc. of Farm Managers and Rural Appraisers. 1977 (2nd Ed.). Professional Farm-ranch Management Manual. Denver, Colo. 200 p.

Designed for professional farm and ranch managers; farm management services, evaluating the farm business, leases and contracts, and farm and ranch planning and reporting.

_____. 1975 (4th Ed.). Professional Rural Appraisal Manual. Stipes Pub. Co., Champaign, Ill. 263 p.

A manual outlining the ethics, commonly used methods, and principals of rural appraising; includes appraisal exhibits.

Bonnen, C.A., and J.M. Ward. 1955. Some Economic Effects of Drought on Ranch Resources. Texas Agric. Expt. Sta. Bul. 801. 11 p.

A study of effects of drought on livestock numbers, range conditions, financial resources, and management of ranches in western Texas.

Boykin, Calvin C. 1968. Economic and Operational Characteristics of Cattle Ranches: Texas High Plains and Rolling Plains. Texas Agric. Expt. Sta. Misc. Pub. 866. 20 p.

A study of the characteristics, investment, costs and income, and factors affecting ranch income with common ranch resource situations.

Boykin, Calvin C., Douglas D. Caton, and Lynn Rader. 1966. Economic and Operational Characteristics of Arizona and New Mexico Range Cattle Ranches. USDA, Econ. Res. Serv. ERS-260. 25 p.

Presents the costs, income, and investment of typical ranches in the southern Intermountain and southern desert ranch areas.

Boykin, Calvin C., and Nathan K. Forrest. 1971. Economic and Operational Characteristics of Livestock Ranches--Edwards Plateau and Central Basin of Texas. Texas Agric. Expt. Sta. Misc. Pub. 978. 30 p.

A study of the characteristics, investment, costs and income, and factors affecting ranch income with common ranch resource situations.

Boykin, C[alvin].C., N[athan].K. Forrest, and John Adams. 1972. Economic and Operational Characteristics of Livestock Ranches--Rio Grande Plains and Trans-Pecos of Texas. Texas Agric. Expt. Sta. Misc. Pub. 1055. 27 p.

A study of the characteristics, investment, costs and income, and factors affecting ranch income with common ranch resource situations.

Carver, R.D., and G.A. Helmers. 1975. Growth Potential of Sandhills Ranches Through Irrigation. Neb. Agric. Expt. Sta. Res. Bul. 266. 55 p.

Comparison of extensive growth of small and medium-sized Nebraska Sandhills ranches through land purchase or rental with intensive growth through irrigation.

Caton, Douglas D. 1965. Effects of Changes in Grazing Fees and Permitted Use of Public Rangelands on Incomes of Western Livestock Ranches. USDA ERS-248. 33 p.

A study of the economic effects of alternative levels of public land grazing fees and grazing permits.

Cook, C. Wayne, E.T. Bartlett, and Gary R. Evans. 1974. A Systems Approach to Range Beef Production. Colo. State Univ., Range Sci. Dept. Sci. Ser. 15. 70 p.

Demonstration of a systems analysis approach using computer programs as an aid in selecting appropriate alternatives in managing a ranching enterprise.

Cornelius, Grant L. 1964. Comparison of Alternative Beef Cattle Systems for Western South Dakota Ranches. S. Dak. Agric. Expt. Sta. Agric. Econ. Pamphlet 117. 31 p.

Study in western South Dakota to determine net ranch income from different sizes and types of commercial cattle ranch operations.

Dean, G.W., A.J. Finch, and J.A. Petit, Jr. 1966. Economic Strategies for Foothill Beef Cattle Ranchers. Calif. Agric. Expt. Sta. Bul. 824. 48 p.

Analysis of the economics of several beef cattle production systems superimposed on the actual resources of a Sacramento valley foothill ranch.

Eisgruber, L., G. Nelson, M. Becker, G. Blanch, et al. 1975. The Cowman's Management Options for 1975 and Beyond. Ore. Agric. Expt. Sta. Spec. Rep. 442. 25 p.

Presents a brief statement of the outlook situation and a number of management strategies to aid cattlemen in arriving at the best management strategy for their operations.

Epp, A.W., and Robert E. Perry. c1974. The Sandhills Ranch Business in 1970. Neb. Agric. Expt. Sta. Bul. 525. 11 p.

Comparisons by ranch size groups of investment, costs and returns, and factors influencing earnings of Sandhills ranches for 1960, 1965, and 1970.

Furniss, I.F., and V.W. Yorgason. 1971 (Rev.). The Economics of Beef Production. Can. Dept. Agric. Pub. 1356. 33 p.

With emphasis on trends in the beef industry and on the cow-calf business, beef cattle finishing, dairy beef, and marketing channels.

Gee, C.Kerry. 1972. Economic and Operational Characteristics of Colorado Range Cattle Business. Colo. Agric. Expt. Sta. Bul. 550. 48 p.

Emphasis given to business ownership and management, resource organization, production practices, livestock marketing, and land management.

Gee, C. Kerry, and Jerry K. Pursley. 1972. The Economics of Retained Ownership of Calves on Eastern Colorado Cattle Ranches. Colo. Agric. Expt. Sta. Bul. 551. 13 p.

Income potential of eastern Colorado plains ranches from selling beef calves as weaners, short yearlings, long yearlings, or as fat cattle.

Gee, C. Kerry, and Melvin D. Skold. 1970. Optimum Enterprise Combinations and Resource Use on Mountain Cattle Ranches in Colorado. Colo. Agric. Expt. Sta. Bul. 546. 12 p.

An analysis of alternative ranch organizations that may increase returns to ranch businesses in the mountain areas of Colorado.

Gee, C. Kerry, and Richard S. Magleby. 1976. Characteristics of Sheep Production in the Western United States. USDA, Agric. Econ. Rep. 345. 47 p.

Current structural characteristics and operational practices in western sheep production.

Goodsell, Wylie D., James R. Gray, and Macie J. Belfield. 1974. Southwest Cattle Ranches, Organization, Costs, and Returns, 1964-72. USDA Agric. Econ. Rep. 255. 37 p.

Operational characteristics, organization, and returns of southwestern cattle ranches, 1964-72.

Goodsell, Wylie D., and Macie Belfield. 1973. Costs and Returns, Migratory-Sheep Ranches, Utah-Nevada, 1972. USDA ERS-523. 15 p.

Operational characteristics, productivity, and returns on migratory sheep ranches in Utah-Nevada in 1972.

______. 1973. Costs and Returns, Northwest Cattle Ranches, 1972. USDA, Econ. Res. Serv. ERS-525. 9 p.

Operational characteristics, productivity, and returns on northwestern cattle ranches, 1972.

Gray, James R. 1970. Production Practices, Costs, and Returns of Cattle and Sheep Ranches in the Grassland Area of Southeastern New Mexico. N. Mex. Agric. Expt. Sta. Res. Rep. 173. 24 p.

Evaluated cow calf ranches of three sizes and medium-sized sheep ranches; 1965 data.

______. 1970. Production Practices, Costs, and Returns of Cattle Ranches in the Brushland Area of Southwestern New Mexico. N. Mex. Agric. Expt. Sta. Res. Rep. 179. 22 p.

Evaluated large, medium, and small size cattle ranches; 1965 data.

______. 1970. Production Practices, Costs, and Returns of Cattle Ranches in the Mountain and Plateau Area of Northwestern New Mexico. N. Mex. Agric. Expt. Sta. Res. Rep. 174. 23 p.

Evaluated large, medium, and small size cattle ranches; 1965 data.

______. 1971. Organization, Costs, and Incomes of Western Cattle and Sheep Ranches. N. Mex. Agric. Expt. Sta. Bul. 587. 56 p.

Data from small, medium, large, and extra large cow-calf ranches, cow-yearling ranches, yearling ranches, and sheep ranches over 12 western states; 1965 data.

Gray, James R., and J. Ronald Cox. 1971. Decision Theory and Stocking Decisions on New Mexico Cattle Ranches. N. Mex. Agric. Expt. Sta. Res. Rep. 210. 23 p.

Use of a model based on precipitation probabilities to help ranchers make stocking-rate decisions.

Gray, James R., and Larry D. Bedford. 1970. Economic Analysis of Commercial Ranch Recreation Enterprises. N. Mex. Agric. Expt. Sta. Bul. 559. 36 p.

Costs and returns of typical commercial ranch recreation enterprises in New Mexico and their effect upon net ranch income.

Jameson, Donald A., Sandy A. D'Aquino, and E.T. Bartlett. 1974. Economics and Management Planning of Range Ecosystems. A.A. Balkema Pub., Rotterdam, Netherlands. 244 p.

Presents the theory and research dealing with various natural resource operations and the consequences and techniques of resource management decisions.

Kearl, W. Gordon. 1969. Comparative Livestock Systems for Wyoming Northern Plains Cattle Ranching. Wyo. Agric. Expt. Sta. Bul. 504. 42 p.

Comparison of level of income and variability in income of different cattle production systems.

_____. 1972. Economic Comparisons of the Cow-calf and Cow-yearling Systems for Northern Plains Cattle Ranching. Wyo. Agric. Expt. Sta. Res. J. 67. 27 p.

Included demonstration of the effects of variations in prices, percent calf crop, and weaning weight of calves on comparative net ranch incomes.

Lovering, James, and David MacMinn. 1973. Estimating Costs of Beef Production. Can. Dept. Agric. Pub. 1506. 53 p.

Estimates and procedures for estimating production costs in cow-calf and feeder enterprises.

Maddox, L.A., Jr. 1973. Management Controls for Ranches Producing Breeding Cattle. Texas Agric. Ext. Bul. 1145. 19 p.

Record keeping system to provide useful management control information.

_____. 1974. Management Controls for Large Ranches. Texas Agric. Ext. Bul. 1142. 13 p.

Record keeping system to provide useful management control information.

______. 1974. Management Controls for Small Ranches. Texas Agric. Ext. Bul. 1143. 9 p.

Record keeping system to provide useful management control information.

Meuller, R.G. 1968. Costs of Cow-calf Ranching in Northern Interior Washington. Wash. Agric. Expt. Sta. Cir. 481. 22 p.

Gives special consideration to season of calving and ranch size.

Mitchell, Burke, and James R. Garrett. 1977. Characteristics of the Range Cattle Industry, 1972. Region III, Northeastern Nevada. Nev. Agric. Expt. Sta. Bul. B42. 15 p.

Costs and returns and management practices of range cow-calf and cow-yearling operations in northeastern Nevada.

Neely, W.V. 1963. A Management Tool for Range Evaluation. Nev. Agric. Ext. Bul. 111. 14 p.

Economic range evaluation by (1) rental rates, (2) values in terms of alternative feeds, and (3) values in terms of animal productivity.

Nelson, C. Alan, and Melvin D. Skold. 1970. Resources, Costs and Returns on Cattle Ranches in the Mountain Areas of Colorado by Size of Ranch. Colo. Agric. Expt. Sta. Tech. Bul. 101. 45 p.

Study ranch size varied from 74 to 764 animal units.

Ott, Gene. 1973. Easier Decisions With Partial Budgeting. N. Mex. Agric. Ext. Cir. 452. 13 p.

Preparation and use of partial budgets in making sound farm and ranch decisions.

Peryam, J. Stephen, and Carl. E. Olson. 1975. Impact of Potential Changes in BLM Grazing Policies on West-central Wyoming Cattle Ranches. Wyo. Agric. Expt. Sta. Res. J. 87. 16 p.

Effects of increased grazing fees and decreased permit numbers on returns to ranch operator's labor, management, and owned capital and on ranch resource organization.

Quenemoen, M.E., and Layton S. Thompson. 1965. How to Estimate Land Value. Mon. Agric. Expt. Sta. Bul. 327. 27 p.

Methods useful in evaluating land and factors affecting land prices.

Reed, A.D., and L.A. Horel. 1976 (Rev.). Beef-planning Profitable Production. Calif. Agric. Ext. Leaflet 2320. 20 p.

Preplanning and budgeting for determining the best system and size of beef production to use the available feed on a ranch.

Rogers, LeRoy F. 1965. Characteristics of the Range Cattle Industry in Nevada. Region I. Southern Nevada. Nev. Agric. Expt. Sta. Bul. B-5. 28 p.

Investment, ranch receipts and expenses, and management practices on cattle ranches in southern Nevada; 1963 basis.

Rogers, LeRoy F., and William C. Helming. 1966. Characteristics of the Range Cattle Industry in Nevada. Region II. Western Nevada. Nev. Agric. Expt. Sta. Bul. B-8. 38 p.

Investment, ranch receipts and expenses, and management practices on cattle ranches in western Nevada; 1963 basis.

______. 1967. Characteristics of the Range Cattle Industry in Nevada. Region III. Northeastern Nevada. Nev. Agric. Expt. Sta. Bul. B-11. 24 p.

Investment, ranch receipts and expenses, and management practices on cattle ranches in northeastern Nevada; 1963 basis.

Shultis, Arthur. 1962. Estimating Rangeland Values and Rents. Calif. Agric. Ext. Serv. Pub. AXT-70. 7 p.

Estimating a fair price for pasturage, rental per acre of range, and agricultural value of range.

Stevens, Delwin M. 1971. An Economic Analysis of Wyoming's Sheep Industry (1960, 1964, 1968). Wyo. Agric. Expt. Sta. Bul. 546. 64 p.

Economic aspects of range and farm flock sheep production in Wyoming; provides economic data and information on management practices for making management decisions.

______. 1975. Wyoming Mountain Valley Cattle Ranching in 1973 and 1974--An Economic Analysis. Wyo. Agric. Expt. Sta. Res. J. 95. 44 p.

Provides data on ranch organization, livestock management practices, earnings and production costs, and factors influencing earnings for making management decisions.

Stevens, Delwin M., and Daniel R. Hartley. 1976. Decline in Wyoming Sheep Industry--A Partial Explanation. Wyo. Agric. Expt. Sta. Res. J. 104. 41 p.

Factors causing termination or reduction of sheep enterprises, and current uses of resources taken out of sheep production.

USDA and USDI. 1977. Study of Fees for Grazing Livestock on Federal Lands. U.S. Govt. Print. Office, Washington, D.C. Paging varies.

History of grazing fees, discussion of issues related to fees, presentation of alternative fee determination procedures, and the Secretaries' recommendations and conclusions.

Vallentine, John F., Donald C. Clanton, Donald F. Burzlaff, and Paul Q. Guyer. 1964. Your Cattle Ranch Business. Neb. Agric. Ext. Cir. 64-211. 12 p.

Analyzing the ranch business, improving ranch production, effective ranch organization, and economics of range management and improvement.

Wendland, Kenneth H., and James R. Gray. 1968. Economic Aspects of Registered Cattle Enterprises in New Mexico. N. Mex. Agric. Expt. Sta. Bul. 538. 25 p.

Operational practices, investment, and costs and returns of registered cattle enterprises in New Mexico.

Range Research and Education

Amer. Soc. Range Mgt., Comm. for Coop. with Youth Org. 1961 (Rev.). Range, Its Nature and Use. Amer. Soc. Range Mgt., Portland, Ore. 65 p.

A manual for youth groups prepared by a youth committee of the American Society of Range Management.

Brown, Dorothy. 1954. Methods of Surveying and Measuring Vegetation. Commonwealth Agric. Bur. Pastures and Field Crops Bul. 42. 223 p.

A manual for quantitative ecological studies of vegetation; methods are classified under botanical analysis, productivity, and utilization and are those applicable primarily to grazing lands.

Campbell, J. Baden (Comp.). 1969. Experimental Methods for Evaluating Herbage. Can. Dept. of Agric. Pub. 1315. 223 p.

Experimental procedures and analytical techniques for measurement and evaluation of forages; chapters were contributed by various subject specialists.

Golley, Frank B., and Helmut K. Buechner (Eds.). 1968. A Practical Guide to the Study of the Productivity of Large Herbivores. Blackwell Sci. Pub., Oxford, Eng. (IBP Handbook 7). 308 p.

Methods of studying ecological and physiological aspects of productivity in wild and domesticated large herbivores.

Harris, L.E., G.P. Lofgreen, C.J. Kercher, R.J. Raleigh, and V.R. Bohman. 1967. Techniques of Research in Range Livestock Nutrition. Utah Agric. Expt. Sta. Bul. 471. 86 p.

An evaluative review of techniques for measuring qualitative and quantitative forage intake of range animals and forage digestibility and for experimentally feeding range livestock.

Joint Committee of Amer. Soc. Agron., Amer. Dairy Sci. Assoc., Amer. Soc. Anim. Prod., and Amer. Soc. Range Mgt. 1962. Pasture and Range Research Techniques. Comstock Pub. Assoc., Ithaca, N.Y. 242 p.

A review of pasture and range research techniques, including both animal and plant phases, with primary emphasis on grazing trials.

Lesperance, A.L., D.C. Clanton, A.B. Nelson, and C.B. Theurer. 1974. Factors Affecting the Apparent Chemical Composition of Fistula Samples. Nev. Agric. Expt. Sta. Tech. Bul. T18.

Considerations of problems of losses during collection, animal contamination, and chemical changes resulting from sample preparation in using fistula techniques.

Natl. Res. Council, Subcomm. on Range Res. Methods. 1962. Basic Problems and Techniques in Range Research. Natl. Acad. Sci.-Natl. Res. Counc. Pub. 890. 341 p.

Range animal, plant, and land research methods and their uses, limitations, and suitability in range research.

USDA, For. Serv. 1959. Techniques and Methods of Measuring Understory Vegetation. USDA, Southern For. Expt. Sta. and Southeastern For. Expt. Sta. (Proceedings of Symposium at Tifton, Georgia, Oct. 1958). 174 p.

Techniques of measuring herbage production and utilization and plant cover and composition along with a consideration of special problems in measurement.

_____. 1962. Range Research Methods: A Symposium. USDA Misc. Pub. 940. 172 p.

A symposium of papers giving special emphasis to vegetation measurement and sampling, site evaluation, measurement of range utilization, and the design and conduct of grazing experiments.

Wright, Madison J. (Ed.). 1973. Range Research and Range Problems. Crop Sci. Soc. Amer. Spec. Pub. 3. 91 p.

Papers of a symposium; range problems and the application of the results of range research to their solution.

INTRODUCTORY BIBLIOGRAPHY OF FOREIGN RANGE MANAGEMENT[50]

Foreign (General)

Anderson, Ray. 1977. Nomads and the Rangeland. Rangeman's J. 4(1):5-6.

Bogdan, A.V. 1977. Tropical Pasture and Fodder Plants. Longman, Inc., New York. 475 p.

Box, Thadis W. 1961. Suggestions for Solving Foreign Range Management Problems. J. Range Mgt. 14(4):179-82.

Corti, Linneo N. 1970. Range Management in the Developing Countries. J. Range Mgt. 23(5):322-24.

Cox, Milo L. 1966. Strengthening Range Management Assistance - The Administrator. J. Range Mgt. 19(6):325-27.

Crisp, D.J. (Ed.). 1964. Grazing in Terrestrial and Marine Environments. Blackwell Sci. Pub., Oxford, Eng. 322 p.

Davies, J.G. 1965. Pasture Improvement in the Tropics. Proc. Internat. Grassland Cong. 9:217-20.

Davies, William, and C.L. Skidmore. 1966. Tropical Pastures. Faber and Faber, London. 215 p.

Doll, E.C., and G.O. Mott (Eds.). 1975. Tropical Forages in Livestock Production Systems. Amer. Soc. Agron. (Madison, Wisc.) Spec. Pub. 24. 104 p.

Elshafie, Mohamed M. 1978. The Significance of Rangelands of Arab States in Animal Production. Proc. Internat. Rangeland Cong. 1:137-39.

French, M.H. 1970. Observations on the Goat. Food and Agric. Organ. of the U.N., Rome, Italy. 204 p.

Hutton, E. Mark. 1978. Humid Tropics--A Sleeping Forage Giant. Proc. Internat. Rangeland Cong. 1:34-36.

Johnston, Alex. 1966. Strengthening Range Management Technical Assistance - The Advisor. J. Range Mgt. 19(6):327-30.

Kelley, Omar J. 1973. AID's Interest in Range Management and Livestock Production in the Tropics and Subtropics. J. Range Mgt. 26(4):242-47.

LeHouerou, H.N., and C.H. Hoste. 1977. Rangeland Production and Annual Rainfall Relations in the Mediterranean Basin and in the African Sahelo-Sudanian Zone. J. Range Mgt. 30(3):181-89.

Pearse, C. Kenneth. 1966. Expanding Horizons in Worldwide Range Management. J. Range Mgt. 19(6):336-40.

Peterson, Roald A. 1964. Improving Technical Assistance in Range Management in Developing Countries. J. Range Mgt. 17(6):305-9.

_____. 1973. The Work of FAO in Range Management. J. Range Mgt. 26(5):316-19.

Pfander, W.H. 1971. Animal Nutrition in the Tropics - Problems and Solutions. J. Anim. Sci. 33(4):843-49.

Puri, G.S. 1966. Grass Resources and Economic Development of Tropical Lands. Proc. Internat. Grassland Cong. 10:818-23.

Riley, Denis, and Anthony Young. 1968. World Vegetation. Cambridge Univ. Press, London. 96 p.

Riney, Thane. 1966. Improvement of Range Management on Forest Lands. Proc. Internat. Grassland Cong. 10:813-17.

Semple, A.T. 1970. Grassland Improvement. CRC Press, Cleveland, Ohio. 400 p.

Shaw, N.H., and W.W. Bryan (Eds.). 1976. Tropical Pasture Research Principles and Methods. Commonwealth Agric. Bur. (Farnham Royal, Bucks, Eng.). Bul. 51. 454 p.

Spedding, C.R.W. 1971. Grassland Ecology. Clarendon Press, Oxford, Eng. 221 p.

Stewart, Philip J. 1978. Islamaic Law as a Factor in Grazing Management: The Pilgrimage Sacrifice. Proc. Internat. Rangeland Cong. 1:119-20.

Stoddart, Laurence A., Arthur D. Smith, and Thadis W. Box. 1975. (3rd Ed.). Grazing Areas of the World. Chapter 2. In Range Management. McGraw-Hill Book Co., New York.

Tomanek, G.W. 1966. How the Society Can Help Individual Advisors and Country Workers. J. Range Mgt. 19(6):333-35.

UNESCO, Standing Comm. on Classification and Mapping of Vegetation on a World Basis. 1973. International Classification and Mapping of Vegetation. UNESCO, Paris, France. 93 p.

Walter, Heinrich (D. Mueller-Dombois, Trans.; J.H. Burnett, Ed.). 1971. Ecology of Tropical and Subtropical Vegetation. Oliver and Boyd, Edinburgh, Scotland. 539 p.

Whyte, R[obert].O. 1968. Grasslands of the Monsoon. Frederick A. Prager, Pub., New York. 325 p.

_____. 1974. Tropical Grazing Lands: Communities and Constituent Species. Dr. W. Junk b.v., Pub., The Hague, The Neth. 220 p.

Williams, Robert E., B.W. Allred, Reginald M. Denio, and Harold A. Paulsen, Jr. 1968. Conservation, Development, and Use of the World's Rangelands. J. Range Mgt. 21(6):355-60.

Australasia

Addison, K.B. 1970. Management Systems on Spear Grass Country. Proc. Internat. Grassland Cong. 11:789-93.

Alexander, G., and O.B. Williams (Eds.). 1973. The Pastoral Industries of Australia. Sydney University Press, Sydney, Austr. 567 p.

Alexander, G.I., J.J. Daly, and M.A. Burns. 1970. Nitrogen and Energy Supplements for Grazing Beef Cattle. Proc. Internat. Grassland Cong. 11:793-96.

Anderson, Kling L. 1954. Grassland Management in New Zealand. J. Range Mgt. 7(4):155-61.

Arnold, G.W. 1977. Effects of Herbivores on Arid and Semi-arid Rangelands: Defoliation and Growth of Forage Plants. Proc. 2nd. United States/Australia Rangeland Panel, Adelaide, 1972:57-73 (Austr. Rangeland Soc., Perth, W. Austr.).

Barker, Susan, and R.T. Lange. 1969. Effects of Moderate Sheep Stocking on Plant Populations of a Black Oak-Bluebush Association. Aust. J. Bot. 17(3): 527-37.

Beale, I.F. 1973. Tree Density Effects on Yields of Herbage and Tree Components in South-west Queensland Mulga (_Acacia aneura_ F. Muell.) Scrub. Trop. Grasslands 7:135-42.

Beattie, W.A. 1956. Beef-cattle Industry of Australia. CSIRO Bul. 278. 135 p.

Box, Thadis W., and Rayden A. Perry. 1971. Rangeland Management in Australia. J. Range Mgt. 24(3):167-71.

Boyland, D.E. 1973. Vegetation of Mulga Lands With Special Reference to South-western Queensland. Trop. Grasslands 7:35-42.

Britten, E.J. 1962. Evaluation for Pasture Purposes of Some African Clovers in a Plant Introduction Program. J. Range Mgt. 15(6):329-33.

Brown, T.H. 1970. The Effect of Autumn Saving of Pasture on Wool Production in a Mediterranean Environment in Southern Australia. Proc. Internat. Grassland Cong. 11:869-73.

Burrows, W.H., and I.F. Beale. 1969. Structure and Association in the Mulga (Acacia aneura) Lands of South-western Queensland. Aust. J. Bot. 17(3):539-52.

_____. 1970. Dimension and Production Relations of Mulga (Acacia aneura F. Muell.) Trees in Semi-arid Queensland. Proc. Internat. Grassland Cong. 11:33-35.

Cameron, I.D. 1952. New Zealand Hill Country and Its Relation to the Richer Pastures of the Lower Lands. Proc. Internat. Grassland Cong. 6:556-61.

Cameron, I.H., and D.J. Cannon. 1970. Changes in the Botanical Composition of Pasture in Relation to Rate of Stocking With Sheep, and Consequent Effects on Wool Production. Proc. Internat. Grassland Cong. 11:640-43.

Cayley, J.W.D., A.H. Bishop, and H.A. Birrell. 1974. The Role of Perennial Grass Species in the Western District of Victoria. Proc. 12th Intern. Grassland Cong., Vol. 3, Pt. 1, p. 97-104.

Chapline, W.R. 1971. Range Management and Improvement in New Zealand. J. Range Mgt. 24(5):329-33.

Christian, C.S. 1952. Cattle Pastures of Tropical Australia. Proc. Internat. Grassland Cong. 6:1534-39.

Christie, E.K. 1975. Physiological Responses of Semi-arid Grasses. III. Growth in Relation to Temperature and Soil Water Deficit. Aust. J. Agric. Res. 26(3):447-57.

Chudleigh, P.D., and S.J. Filan. 1972. A Simulation Model of an Arid Zone Sheep Property. Aust. J. Agric. Econ. 16:183-94.

Cowling, S.W. 1977. Effects of Herbivores on Nutrient Cycling and Distribution in Rangeland Ecosystems. Proc. 2nd. United States/Australia Rangeland Panel, Adelaide, 1972:277-99 (Austr. Rangeland Soc., Perth, W. Austr.).

C.S.I.R.O. (Brisbane, Austr.). 1964. Some Concepts and methods in Subtropical Pasture Research. Commonwealth Agricultural Bureau, Farnham Royal, Bucks, Eng. 242 p.

Cunningham, G.M. 1976. Range Management in Western New South Wales. Wool Tech. Sheep Breed. 23:21-30.

Davies, J. Griffiths. 1960. Pasture and Forage Legumes for the Dry Subtropics and Tropics of Australia. Proc. Internat. Grassland Cong. 8:381-85.

Davies, J.J.F. 1975. Land Use by Emus and Other Wildlife Species in the Arid Shrublands of Western Australia. Arid Shrublands--Proceedings of the Third Workshop of the United States/Australia Rangelands Panel, 1973:91-98.

Davies, S.J.J.F. 1973. Environmental Variables and the Biology of Native Australian Animals of the Mulga Lands. Trop. Grasslands 7:127-34.

Dawson, N.M., D.E. Boyland, and C.R. Ahern. 1975. Land Management in South-west Queensland. Proc. Ecol. Soc. Aust. 9:124-41.

Dawson, Roy C. 1962. Inoculation for Better Pasture and Forage Legumes in the Tropics. J. Range Mgt. 15(5):252-57.

Dunbar, Graham A. 1978. Regeneration of High Mountain Rangeland Sites in New Zealand After Cultural Treatment. Proc. Internat. Rangeland Cong. 1: 712-14.

Easter, C.D. 1975. Some Agronomic Aspects of Factors Underlying Production in New South Wales Grazing Industry. Q. Rev. Agric. Econ. 28(3):177-99.

Ebersohn, J.P. 1970. Development of a Pastoral Holding in Semiarid Central Western Australia. Proc. Internat. Grassland Cong. 11:175-79.

Evans, T.R. 1970. Some Factors Affecting Beef Production from Queensland. Proc. Internat. Grassland Cong. 11:803-7.

Everett, Rodney A., and F. James Vickery. 1978. Commercial Horse Breeding as a Potential Alternative Grazing Use of South Australian Arid Rangelands. Proc. Internat. Rangeland Cong. 1:588-91.

Everist, S[elwyn].L. 1969. Use of Fodder Trees and Shrubs. Queensland Div. Plant Ind. Advisory Leaflet 1024. 44 p.

_____. 1972. Continental Aspects of Shrub Distribution, Utilization, and Potentials: Australia. USDA, For. Serv. Gen. Tech. Rep. INT-1, p. 16-25.

Falvey, J. Lindsay. 1977. Dry Season Regrowth of Six Forage Species Following Wildfire. J. Range Mgt. 30(1):37-39.

Fels, H.E., R.J. Moir, and R.C. Rossiter. 1959. Herbage Intake of Grazing Sheep in South-western Australia. Austr. J. Agric. Res. 10(2):237-47.

Fitzgerald, K. 1975. Regional Pasture Development and Associated Problems--Northern Western Australia. Trop. Grasslands 9:77-92.

Gillard, P. 1970. Pasture Development in the Dry Tropics of North Queensland. Proc. Internat. Grassland Cong. 11:807-10.

Goodall, D.W. 1971. Extensive Grazing Systems. Pages 173-87 in Systems Analysis in Agricultural Management. John Wiley and Sons, Sydney, Austr.

Graetz, R.D. 1975. Biological Characteristics of Australian Acacia and Chenopodiaceous Shrublands Relevant to Their Pastoral Use. Arid Shrublands--Proceedings of the Third Workshop of the United States/Australia Rangelands Panel, 1973:33-39.

Griffiths, M., R. Barker, and L. MacLean. 1974. Further Observations on the Plants Eaten by Kangaroos and Sheep Grazing Together in a Paddock in South-western Queensland. Aust. Wildl. Res. 1:27-43.

Hagon, M.W., and R.H. Groves. 1977. Some Factors Affecting the Establishment of Four Native Grasses. Austr. J. Expt. Agric. and Anim. Husb. 17(84):90-96.

Harrington, Graham. 1978. The Implications of Goat, Sheep and Cattle Diets to the Management of an Australian Semiarid Woodland. Proc. Internat. Rangeland Cong. 1:447-50.

Heady, Harold F. 1967. Practices in Range Forage Production. Univ. Queensland Press, St. Lucia, Queensland, Austr. 81 p.

Howes, K.M.W. (Ed.). 1978. Studies of the Australian Arid Zone. III. Water in Rangelands. CSIRO, Melbourne, Aust.

Hutton, E.M. 1970. Australian Research in Pasture Plant Introduction and Breeding. Proc. Internat. Grassland Cong. 11:A1-A12.

Johns, A.T. 1955. Pasture Quality and Ruminant Digestion. I. Seasonal Change in Botanical and Chemical Composition of Pasture. New Zeal. J. Sci. and Tech. 37(4):302-11.

Jozwik, F.X., A.O. Nicholls, and R.A. Perry. 1970. Studies on the Mitchell Grasses (Astrebla F. Muell.). Proc. Internat. Grassland Cong. 11:48-51.

Lange, R.T. 1977. The Nature of Arid and Semi-arid Ecosystems as Rangeland. Proc. 2nd. United States/Australia Rangeland Panel, Adelaide, 1972: 15-27 (Austr. Rangeland Soc., Perth, W. Austr.).

Langlands, F.P., and J.E. Bowles. 1976. Nitrogen Supplementation of Ruminants Grazing Native Pastures in New England, New South Wales. Austr. J. Expt. Agric. and Anim. Husb. 16(82):630-35.

Leigh, J.H. 1977. Some Effects of Diet Selectivity by Vertebrate and Invertebrate Herbivores on Arid and Semi-arid Rangelands. Proc. 2nd. United States/Australia Rangeland Panel, Adelaide, 1972:125-49. (Austr. Rangeland Soc., Perth, W. Austr.).

Leigh, J.H., A.D. Wilson, and O.B. Williams. 1970. An Assessment of the Value of Three Perennial Chenopodiaceous Shrubs for Wool Production of Sheep Grazing Semi-arid Pastures. Proc. Internat. Grassland Cong. 11:55-59.

Leigh, J.H., and W.E. Mulham. 1971. The Effect of Defoliation on the Persistence of Atriplex vesicaria. Aust. J. Agric. Res. 22(2):239-44.

Lendon, C., and R.R. Lamacraft. 1976. Standards for Testing and Assessing Range Condition in Central Australia. Aust. Rangeland J. 1(1):40-48.

Loneragan, J.F. 1970. The Contribution of Research in Plant Nutrition to the Development of Australian Pastures. Proc. Internat. Grassland Cong. 11:A13-A22.

Low, W.A. 1977. Behavior of Herbivores (Except Sheep) Influencing Rangelands in Australia. Proc. 2nd. United States/Australia Rangeland Panel, Adelaide, 1972:149-63 (Austr. Rangeland Soc., Perth, W. Austr.).

Luck, P.E. 1970. The Role of Improved Pastures in the Near North Coast Region of Southeast Queensland. Proc. Internat. Grassland Cong. 11:161-65.

Lynch, J.J. 1977. Movement of Some Rangeland Herbivores in Relation to Their Feed and Water Supply. Proc. 2nd. United States/Australia Rangeland Panel, Adelaide, 1972:163-73 (Austr. Rangeland Soc., Perth, W. Austr.).

McCarron, A.C. 1975. The Changing Structure and Economic Situation of the Australian Sheep and Beef Cattle Grazing Industry. Q. Rev. Agric. Econ. 28(3):152-76.

McKeown, N.R., and R.C.G. Smith. 1970. Seasonal Pasture Production, Liveweight Change and Wool Growth of Sheep in a Mediterranean Environment. Proc. Internat. Grassland Cong. 11:873-76.

McMeekan, C.P. 1960. Grazing Management. Proc. Internat. Grassland Cong. 8:21-27.

Malcolm, C.V. 1972. Establishing Shrubs in Saline Environments. USDA, For. Serv. Gen. Tech. Rep. INT-1, p. 392-403.

Michalk, D.L. 1978. Range Management in Australia. Rangeman's J. 5(5): 143-45.

Michalk, D.L., and J.A. Beale. 1976. An Evaluation of Barrel Medic (*Medicago truncatula*) as an Introduced Pasture Legume for Marginal Cropping Areas of Southeastern Australia. J. Range Mgt. 29(4):328-33.

Michalk, D.L., C.C. Byrnes, and G.E. Robards. 1976. Effects of Grazing Management on Natural Pastures in a Marginal Area of Southeastern Australia. J. Range Mgt. 29(5):380-83.

Minson, D.J., and M.N. McLeod. 1970. The Digestibility of Temperate and Tropical Grasses. Proc. Internat. Grassland Cong. 11:719-23.

Moore, R.M[ilton]. 1965. Ecological Effects of Grazing on Grasslands in Southeastern Australia. Proc. Internat. Grassland Cong. 9:429-33.

_____. 1970. Australian Grasslands. Australian Natl. Univ. Press, Canberra. 455 p.

_____. 1975. Australian Arid Shrublands. Arid Shrublands--Proceedings of the Third Workshop of the United States/Australian Rangelands Panel, 1973:6-11.

Moore, R. Milton, and Joseph Walker. 1974. Interrelationships Among Trees, Shrubs, and Herbs in Australian Shrub Woodlands. Proc. 12th. Intern. Grassland Cong., Vol. 1, Pt. 2, p. 756-62.

Moore, R.M., Nancy Barrie, E.H. Kipps, G.A. McIntyre, et al. 1954. Grazing Management: Continuous and Rotation Grazing. CSIRO Bul. 201. 104 p.

Mott, J.J., and A.J. McComb. 1975. Effects of Moisture Stress on the Growth and Reproduction of Three Annual Species From an Arid Region of Western Australia. J. Ecol. 63:825-34.

Murray, R.M., C. Graham, J. Round, and J. Bond. 1978. Intake and Response to Phosphorus Supplements by Range Cattle. Proc. Internat. Rangeland Cong. 1:427-28.

Newman, J.C. 1974. Effects of Past Grazing in Determining Range Management Principles in Australia. USDA Misc. Pub. 1271, p. 197-206.

Newman, R.J. 1966. Problems of Grassland Establishment and Maintenance on Hill-country in Victoria. Proc. Internat. Grassland Cong. 10:875-78.

Newsome, R.E., and L.K. Corbett. 1977. The Effects of Native, Feral, and Domestic Animals on the Productivity of Australian Rangelands. Proc. 2nd. United States/Australia Rangeland Panel, Adelaide, 1972:331-57 (Austr. Rangeland Soc., Perth, W. Austr.).

O'Connor, Kevin F., and I.G. Christopher Kerr. 1978. The History and Present Pattern of Pastoral Range Production in New Zealand. Proc. Internat. Rangeland Cong. 1:104-7.

Orr, D.M. 1975. A Review of Astrebla (Mitchell Grass) Pastures in Australia. Trop. Grasslands 9(1):21-36.

Perry, R.A. 1970. The Effects on Grass and Browse Production of Various Treatments on a Mulga Community in Central Australia. Proc. Internat. Grassland Cong. 11:63-66.

______. 1975. Range Sites in Australia. Arid Shrublands--Proceedings of the Third Workshop of the United States/Australia Rangelands Panel, 1973:23-25.

______. 1977. Rangeland Management for Livestock Production in Semiarid and Arid Australia. Proc. 2nd. United States/Australia Rangeland Panel, Adelaide, 1972:311-17 (Austr. Rangeland Soc., Perth, W. Austr.).

Philipp, Perry F. 1959. The Economics of Grassland Development and Improvement in New Zealand. J. Range Mgt. 12(4):170-75.

Pressland, A.J. 1975. Productivity and Management of Mulga in South Western Queensland Related to Tree Structure and Density. Aust. J. Bot. 23(6): 965-76.

______. 1976. Possible Effects of Removal of Mulga on Rangeland Stability in South Western Queensland. Aust. Rangeland J. 1(1):24-30.

Purcell, D.L., and G.R. Lee. 1970. Effects of Season and of Burning Plus Planned Stocking on Mitchell Grass Grasslands in Central Western Queensland. Proc. Internat. Grassland Cong. 11:66-69.

Ritchie, R.B., and A.H. Bishop. 1974. The Management of a Grazing Property in Southern Australia. Proc. 12th. Intern. Grassland Cong., Vol. 3, Pt. 1, p. 438-44.

Rixon, A.J. 1971. Oxygen Uptake and Nitrification by Soil Within a Grazed *Atriplex vesicaria* Community in Semiarid Rangeland. J. Range Mgt. 24(6):435-39.

Roe, R., W.H. Southcott, and Helen Newton Turner. 1959. Grazing Management of Native Pastures in the New England Region of New South Wales. I. Pasture and Sheep Production with Special Reference to Systems of Grazing and Internal Parasites. Austr. J. Agric. Res. 10(4):530-54.

Sadler, B.S. 1976. Water Resource and Land Use Problems in Western Australia. Watershed Management on Range and Forest Lands, Proceedings of the Fifth Workshop of the United States/Australia Rangelands Panel, 1975:13-30.

Scateni, W.J. 1966. Effect of Variations in Stocking Rate and Conservation on the Productivity of Sub-tropical Pastures. Proc. Internat. Grassland Cong. 10:947-51.

Schinckel, P.G. 1960. Variation in Feed Intake as a Cause of Variation in Wool Production of Grazing Sheep. Austr. J. Agric. Res. 11(4):585-94.

Schmidl, Laszlo, and Keith Turnbull. 1977. Ragwort (*Senecio jacobaea* L.) and Blackberry (*Rubus fruticosus* L. agg.) in Grasslands of Victoria, Australia. Proc. 13th. Intern. Grassland Cong., Sect. 8-10, p. 61-66.

Scott, D. 1974. Interaction Between Resident Vegetation and Oversown Seed in Improving New Zealand Tussock Grasslands. Proc. 12th. Intern. Grassland Cong., Vol. I, Pt. 2, p. 831-40.

Shaw, N.H. 1970. The Choice of Stocking Rate Treatments as Influenced by the Expression of Stocking Rate. Proc. Internat. Grassland Cong. 11:909-13.

Slatyer, R.O. 1975. Structure and Function of Australian Arid Shrublands. Arid Shrublands--Proceedings of the Third Workshop of the United States/Australia Rangelands Panel, 1973:66-73.

Slatyer, R.O., and R.A. Perry (Eds.). 1970. Arid Lands of Australia. Australian Natl. Univ. Press, Canberra. 321 p.

Smith, C.A. 1970. The Feeding Value of Tropical Grass Pastures Evaluated by Cattle Weight Gains. Proc. Internat. Grassland Cong. 11:839-42.

Smith, D.F. 1965. The Instability of Annual Pastures in Southern Australia. Proc. Internat. Grassland Cong. 9:421-24.

Smith, Edwin L. 1960. Effects of Burning and Clipping at Various Times During the Wet Season on Tropical Tall Grass Range in Northern Australia. J. Range Mgt. 13(4):197-203.

Squires, V[ictor].R. 1970. Grazing Behavior of Sheep in Relation to Watering Points in Semi-arid Rangelands. Proc. Internat. Grassland Cong. 11:880-84.

______. 1976. Walking, Watering and Grazing Behavior of Merino Sheep on Two Semi-arid Rangelands in South-west New South Wales. Aust. Rangeland J. 1(1):13-23.

______. 1978. Distance Trailed to Water and Livestock Response. Proc. Internat. Rangeland Cong. 1:431-34.

Stanley, R.J. 1978. Establishment of Chenopod Shrubs by Tyne Pitting on Hardpan Soils in Western New South Wales, Australia. Proc. Internat. Rangeland Cong. 1:639-42.

Waring, E.J. 1975. Economic and Financial Constraints in the Operation of Livestock Enterprises on Arid Shrublands. Arid Shrublands--Proceedings of the Third Workshop of the United States/Australia Rangelands Panel, 1973:108-15.

Whittet, J.N. 1952. Essentials Underlying Selection of Species for Range and Other Dry-area Zone Reseeding. Proc. Internat. Grassland Cong. 6:521-25.

Williams, O[wen].B. 1970. Longevity and Survival of Some Dietary Constituents in a Natural Semi-arid Grassland Grazed by Sheep. Proc. Internat. Grassland Cong. 11:85-89.

______. 1975. Environmental, Biological, and Managerial Constraints to the Production of Sheep and Cattle in Australian Shrublands. Arid Shrublands--Proceedings of the Third Workshop of the United States/Australia Rangelands Panel, 1973:80-83.

Williams, O[wen].B., and B. Roe. 1975. Management of Arid Grasslands for Sheep: Plant Demography of Six Grasses in Relation to Climate and Grazing. Proc. Ecol. Soc. Aust. 9:142-56.

Williams, Owen B. 1978. Plant Demography of Australian Arid Rangeland and Implications for Management, Research, and Land Policy. Proc. Internat. Rangeland Cong. 1:185-86.

Willoughby, W.M. 1959. Limitations to Animal Production Imposed by Seasonal Fluctuations in Pasture and by Management Procedures. Austr. J. Agric. Res. 10(2):248-68.

Wilson, A.D. 1976. Comparison of Sheep and Cattle Grazing on a Semi-arid Grassland. Austr. J. Agric. Res. 27(1):155-62.

_____. 1977. Grazing Management in the Arid Areas of Australia. Proc. 2nd. United States/Australia Rangeland Panel, Adelaide, 1972:83-93 (Austr. Rangeland Soc., Perth, W. Austr.).

_____. 1977. The Digestibility and Voluntary Intake of the Leaves of Trees and Shrubs by Sheep and Goats. Austr. J. Agric. Res. 28(3):501-8.

_____. 1978. The Relative Production from Sheep, Cattle, and Goats on Temperate Rangelands in Australia. Proc. Internat. Rangeland Cong. 1:444-46.

Wilson, A.D., J.H. Leigh, N.L. Hindley, and W.E. Mulham. 1975. Comparison of the Diets of Goats and Sheep on a *Casuarina cristata* (Belah) - *Heterodendrum oleifolium* (Rosewood) Woodland Community in Western New South Wales. Aust. J. Expt. Agric. and Anim. Husb. 15(72):45-53.

Wilson, Barry (Ed.). 1968. Pasture Improvement in Australia. Murray Pub. Co., Sydney, N.S.W., Austr. 288 p.

Winkworth, R.E. 1963. Some Effects of Furrow Spacing and Depth on Soil Moisture in Central Australia. J. Range Mgt. 16(3):138-42.

_____. 1967. The Composition of Several Arid Spinifex Grasslands of Central Australia in Relation to Rainfall, Soil Water Relations, and Nutrients. Aust. J. Bot. 15(1):107-30.

_____. 1971. Longevity of Buffel Grass Seed Sown in an Arid Australian Range. J. Range Mgt. 24(2):141-45.

Young, J.G., and F. Chippendale. 1970. Beef Cattle Performance on Pastures on Heath Plains in Southeast Queensland. Proc. Internat. Grassland Cong. 11:849-52.

Young, Mike D., and James Vickery. 1978. An Overview of Land Tenure and Administration in Australia's Rangelands. Proc. Internat. Rangeland Cong. 1:97-99.

Zallar, S., and A. Mitchell. 1970. Pasture Species for Non-irrigated Salt-affected Land. Proc. Internat. Grassland Cong. 11:138-42.

Latin America

Aronovich, S., A. Serpa, and H. Ribeiro. 1970. Effect of Nitrogen Fertilizer and Legume Upon Beef Production of Pangolagrass Pasture. Proc. Internat. Grassland Cong. 11:796-800.

Beeson, Kenneth C., and G. Guillermo-Gomez. 1970. Concentration of Nutrients in Pastures in the Central Huallaga and Rio Ucayali Valleys of the Upper Amazon Basin of Peru. Proc. Internat. Grassland Cong. 11:89-92.

Beetle, Alan A. 1954. The Argentine Literature on Range Management. J. Range Mgt. 7(3):125-27.

Bishop, J.P., J.A. Froseth, H.N. Verettoni, and C.H. Noller. 1975. Diet and Performance of Sheep on Rangeland in Semiarid Argentina. J. Range Mgt. 28(1):52-55.

Blydenstein, John. 1956. The Management of Livestock in the Xerophytic Forest Region of Central Argentina. J. Range Mgt. 9(3):113-15.

_____. 1965. The Natural Grasslands of the Colombian Llanos. Proc. Internat. Grassland Cong. 9:1205-6.

_____. 1972. Developing Range Management in Latin America. J. Range Mgt. 25(1):7-9.

Boelcke, Osvaldo, and Roald A. Peterson. 1960. Establishment of New Forage Plants in the Grasslands of Northern Patagonia in Argentina. Proc. Internat. Grassland Cong. 8:159-61.

Buller, Roderic E. 1960. Evaluation of the Forage Resources of the Tropical Grasslands of Mexico. Proc. Internat. Grassland Cong. 8:374-77.

Buller, R[oderic].E., S. Aronovich, L.R. Quinn, and W.V.A. Bisschoff. 1970. Performance of Tropical Legumes in the Upland Savannah of Central Brazil. Proc. Internat. Grassland Cong. 11:143-47.

Butterworth, M.H., and J.A. Diaz L. 1970. Use of Equations to Predict the Nutritive Value of Tropical Grasses. J. Range Mgt. 23(1):55-58.

Chapline, W.R. 1962. Overcoming the Problems of Range Livestock Production in Southern South America. J. Range Mgt. 15(5):259-62.

Chicco, C.F., T.A. Schultz, J. Rios, D. Plasse, and M. Burguera. 1971. Self-feeding Salt-supplement to Grazing Steers Under Tropical Conditions. J. Anim. Sci. 33(1):142-46.

Contreras, David, and Juan Gasto. 1978. Description and Transformation of the Agricultural-Pastoral Ecosystems for the Benefit of Man in Bolivia, Chile, and Peru. Proc. Internat. Rangeland Cong. 1:60-61.

Covas, Guillermo. 1960. Performance of Weeping Lovegrass in the Semiarid Argentine Pampa. Proc. Internat. Grassland Cong. 8:231-34.

Crowder, L.V., H. Chaverra, and J. Lotero. 1970. Productive Improved Grasses in Colombia. Proc. Internat. Grassland Cong. 11:147-49.

Custred, Glynn. 1978. The Utilization of High Altitude Grasslands in the South Central Andes. Proc. Internat. Rangeland Cong. 1:150-52.

DeAndrade, Brenno M. Martins. 1952. Trends on Pasture Establishment and Utilization in Sao Paulo, Brazil. Internat. Grassland Cong. 6:1561-64.

DeAraujo, Anacreonte Avila. 1952. Natural and Artificial Grassland in the State of Parana. Internat. Grassland Cong. 6:1459-63.

Fisher, C.E., and Lawrence Quinn. 1959. Control of Three Major Brush Species on Grazing Lands in the United States, Cuba, and Brazil. J. Range Mgt. 12(5):244-48.

Flores A., Miguel A., and Floyd R. Olive. 1952. Forage Species of El Salvador. Internat. Grassland Cong. 6:1434-39.

Fretes, Ruben A., and Don D. Dwyer. 1969. Range and Livestock Characteristics of Paraguay. J. Range Mgt. 22(5):311-14.

Gonzales, Martin H. 1972. Manipulating Shrub-grass Plant Communities in Arid Zones for Increased Animal Production. USDA, For. Serv. Gen. Tech. Rep. INT-1, p. 429-34.

Graham, Alan (Ed.). 1973. Vegetation and Vegetational History of Northern Latin America. American Elsevier, New York. 393 p.

Grossman, J., S. Aronovich, and E. Do C.B. Campello. 1965. Grasslands of Brazil. Proc. Internat. Grassland Cong. 9:39-47.

Guevara, Juan C., Roberto J. Candia, and Rolando H. Braun W. 1978. Range Resources Inventory of Mendoza (Argentina). Proc. Internat. Rangeland Cong. 1:500-4.

Hermann, Frederick J. 1974. Manual of the Genus Carex in Mexico and Central America. USDA Agric. Handbook 467. 219 p.

Jimenez, Mario Gutierrez. 1952. Forage Plants and Problems in the Highlands of Costa Rica. Proc. Internat. Grassland Cong. 6:1427-33.

Lima, D. DeA. 1965. Vegetation of Brazil. Proc. Internat. Grassland Cong. 9:29-38.

McCorkle, J.S. 1952. Range and Livestock in Tropical Savannahs of British Guiana. J. Range Mgt. 5(4):259-65.

_____. 1968. Ranching in Panama. J. Range Mgt. 21(4):242-47.

McDowell, R.E., and A. Hernandez-Urdaneta. 1975. Intensive Systems for Beef Production in the Tropics. J. Anim. Sci. 41(4):1228-37.

McMahon, P.R. 1966. The Future of Grassland Farming in Argentina and Uruguay. Proc. Internat. Grassland Cong. 10:828-31.

Miles, Wayne. 1977. Ranching in the South American Tropics. Rangeman's J. 4(2):35-36.

Moffat, J.K. 1965. Pasture Establishment in the Beef Cattle Regions of Northern Santa Fe and Corrientes, Argentina. Proc. Internat. Grassland Cong. 9: 287-89.

Morey, Robert V., and Nancy C. Morey. 1978. A Case for Indigenous Participation in Rangeland Development. Proc. Internat. Rangeland Cong. 1:116-18.

Parsons, James J. 1972. Spread of African Pasture Grasses to the American Tropics. J. Range Mgt. 25(1):12-17.

Quinn, L.R.C., G.O. Mott, W.V.A. Bisschoff, and M.B. Jones. 1965. Beef Production of Six Tropical Grasses in Central Brazil. Proc. Internat. Grassland Cong. 9:1015-20.

Richards, J.A. 1970. Productivity of Tropical Pastures in the Caribbean. Proc. Internat. Grassland Cong. 11:A49-A56.

Rogers, Mario A. 1952. Distribution of Cultivated Forage Plants in Chile. Proc. Internat. Grassland Cong. 6:650-54.

Roseveare, G.M. 1948. The Grasslands of Latin America. Imperical Bur. of Pastures and Field Crops (Aberystwyth, Gr. Br.) Bul. 36. 291 p.

Simpson, James R., and Ruben Fretes. 1972. An Economic Evaluation of Buffelgrass in Paraguay. J. Range Mgt. 25(4):261-66.

Soriano, Alberto. 1960. Germination of Twenty Dominant Plants in Patagonia in Relation to Regeneration of the Vegetation. Proc. Internat. Grassland Cong. 8:154-58.

_____. 1972. Continental Aspects of Shrub Distribution Utilization, and Potentials: South America. USDA, For. Serv. Gen. Tech. Rep. INT-1, p. 51-54.

Stonaker, H.H. 1975. Beef Production Systems in the Tropics. I. Extensive Production Systems on Infertile Soils. J. Anim. Sci. 41(4):1218-27.

USDA, Foreign Agric. Serv. 1958. Agricultural Geography of Latin America. USDA Misc. Pub. 743. 96 p.

Veiga, Joao Soares. 1952. Some Problems on Grazing Cattle in Central Brazil. Proc. Internat. Grassland Cong. 6:1520-25.

Venturo, Pedro, and Roberto Alvarez Calderon. 1952. Range Management in the High Andes of Peru. Proc. Internat. Grassland Cong. 6:562-66.

Villares, J.B., A. Tundisi, and M. Beckler. 1953. The Subterranean System of Colonial Grass (Guinea Grass) in Various Soils of the State of Sao Paulo, Brazil. J. Range Mgt. 6(4):248-54.

Whitaker, Morris D., and E. Boyd Wennergren. 1978. Common-Property Rangeland and Overgrazing: Resource Misallocation in Bolivian Agriculture. Proc. Internat. Rangeland Cong. 1:153-55.

White, C. Langdon, and John Thompson. 1955. The Llanos - A Neglected Grazing Resource. J. Range Mgt. 8(1):11-17.

Willard, E. Earl. 1973. Effect of Wildfires on Woody Species in the Monte Region of Argentina. J. Range Mgt. 26(2):97-100.

Woolfolk, E.J. 1955. Range Improvement and Management Problems in Argentina. J. Range Mgt. 8(6):260-54.

Selected Literature of Range Science

Africa

Afolayan, T.A. 1978. Effects of Fire on the Vegetation and Soils in Kainji Lake National Park, Nigeria. Proc. Internat. Rangeland Cong. 1:55-59.

Asare, E.O. 1966. An Experiment in the Use of Aerial Photographs in Range Mapping and Planning in the Coastal Grassland Area of Ghana. Proc. Internat. Grassland Cong. 10:917-20.

Astle, W.L. 1965. The Edaphic Grasslands of Zambia. Proc. Internat. Grassland Cong. 9:363-73.

Ayuko, Lucas J. 1978. Management of Rangelands in Kenya to Increase Beef Production: The Socio-Economic Constraints and Policies. Proc. Internat. Rangeland Cong. 1:82-86.

Bedoian, William H. 1978. Economic Alternatives for a Semipastoral Population in Southeast Tunisia. Proc. Internat. Rangeland Cong. 1:71-75.

Bille, J.C. 1978. Woody Forage Species in the Sahel: Their Biology and Use. Proc. Internat. Rangeland Cong. 1:392-95.

Booysen, Peter de V., and Neil M. Tainton. 1978. Grassland Management: Principles and Practice in South Africa. Proc. Internat. Rangeland Cong. 1: 551-54.

Boudet, G. 1970. Management of Savannah Woodland Range in West Africa. Proc. Internat. Grassland Cong. 11:1-3.

Box, Thadis W. 1968. Range Resources of Somalia. J. Range Mgt. 21(6): 388-92.

Bredemeier, Lorenz F. 1978. Socio-political Practices Hinder Improved Range Management. Proc. Internat. Rangeland Cong. 1:90-91.

Bredon, R.M., D.T. Torell, and B. Marshall. 1967. Measurement of Selective Grazing of Tropical Pastures Using Esophageal Fistulated Steers. J. Range Mgt. 20(5):317-20.

Breman, H., A. Diallo, G. Traore, and M.M. Djiteye. 1978. The Ecology of the Annual Migrations of Cattle in the Shael. Proc. Internat. Rangeland Cong. 1:592-95.

Casebeer, Robert L. 1978. Coordinating Range and Wildlife Management in Kenya. J. For. 76(6):374-75.

Christiansson, Carl. 1978. Relations Between Heavy Grazing, Cultivation, Soil Erosion, and Sedimentation in the Semiarid Parts of Central Tanzania. Proc. Internat. Rangeland Cong. 1:274-78.

Clyburn, Lloyd. 1978. The Process of Change in Certain Livestock Owner and Operating Groups in the West African Sahel. Proc. Internat. Rangeland Cong. 1:108-10.

Codd, L.E.W. 1952. The Results of an Ecological Survey of the Union of South Africa. Proc. Internat. Grassland Cong. 6:596-601.

Darling, F. Fraser. 1960. An Ecological Reconnaissance of the Mara Plains in Kenya Colony. Wildl. Monogr. 5. 41 p.

Davidson, Robert L. 1952. Methods and Procedures for Successful Reseeding of Rangelands in South Africa. Proc. Internat. Grassland Cong. 6:544-47.

_____. 1965. Management of Sown and Natural Lovegrass. J. Range Mgt. 18(4):214-18.

De Fabreques, B.P. 1965. The Study and Principle of the Utilization of Steppe as Pasture in the Republic of Niger. Proc. Internat. Grassland Cong. 9:1433-35.

Denney, Richard N. 1972. Relationships of Wildlife to Livestock on Some Developed Ranches on the Laikipia Plateau, Kenya. J. Range Mgt. 25(6): 415-25.

El Hassan, Babiker A. 1978. Nomadism and Range Management in the Sudan. Proc. Internat. Rangeland Cong. 1:127-29.

Germain, R., and A. Scaut. 1960. Herbage and Nutritional Aspects of Animal Husbandry in the Equatorial Forest of the Congo. Proc. Internat. Grassland Cong. 8:371-74.

Gihad, E.A. 1976. Intake, Digestibility and Nitrogen Utilization of Tropical Natural Grass Hay by Goats and Sheep. J. Anim. Sci. 43(4):879-83.

Goodloe, Sid. 1969. Short Duration Grazing in Rhodesia. J. Range Mgt. 22(6):369-73.

Graves, Walter L., Philip Roark, F. Rudolph Vigil, and Hamidou Bouyayachen. 1978. Increasing Animal Production in Morocco (North Africa) Through Rangeland Renovation and Animal Management. Proc. Internat. Rangeland Cong. 1: 130-32.

Hardman, Billy H., and Leonard Hendzel. 1978. Kenya--Land of Awakening Range Management. Rangeman's J. 5(1):6-8.

Heady, Harold F. 1960. Range Management in the Semi-arid Tropics of East Africa According to Principles Developed in Temperate Climates. Proc. Internat. Grassland Cong. 8:223-26.

Hedberg, Inga, and Olov Hedberg (Eds.). 1968. Conservation of Vegetation in Africa South of the Sahara. Almqvist and Wiksells Boktryckeri AB, Uppsala, Sweden. 320 p.

Helland, Johan. 1978. Sociological Aspects of Pastoral Livestock Production in Africa. Proc. Internat. Rangeland Cong. 1:79-81.

Hickey, Joseph V. 1978. Fulani Nomadism and Herd Maximization: A Model for Government Mixed Farming and Ranching Schemes. Proc. Internat. Rangeland Cong. 1:95-96.

Hildyard, P. 1970. The Utilization of Certain Native Pastures Composed of Grasses of Varying Palatability. Proc. Internat. Grassland Cong. 11:41-45.

Hirst, Stanley M. 1975. Ungulate-habitat Relationships in a South African Woodland/Savanna Ecosystem. Wildl. Monogr. 44. 60 p.

Howell, Denise. 1978. Reclaim Your Veld with Animals. Rangeman's J. 5 (1):3-5.

Howell, L.N. 1978. Development of Multi-camp Grazing Systems in the Southern Orange Free State, Republic of South Africa. J. Range Mgt. 31(6):459-65.

Ibrahim, Kamal M. 1978. Phytogeographical Divisions of Africa. Proc. Internat. Rangeland Cong. 1:177-84.

Ibrahim, Kamal M., and Albert E.O. Chabeda. 1978. Forage Exploration and Evaluation in Kenya. Proc. Internat. Rangeland Cong. 1:339-42.

Kannegieter, A. 1965. The Cultivation of Grasses and Legumes in the Forest Zone of Ghana. Proc. Internat. Grassland Cong. 9:313-18.

Karue, C.N., J.L. Evans, and A.D. Tillman. 1973. Voluntary Intake of Dry Matter by African Zebu Cattle: Quality of Feed and the Reference Base. J. Anim. Sci. 36(6):1181-85.

Larson, Floyd D. 1957. Problems of Population Pressure Upon the Desert Range. J. Range Mgt. 10(4):160-61.

______. 1966. Cultural Conflicts With the Cattle Business in Zambia, Africa. J. Range Mgt. 19(6):367-70.

LeHouerou, H.N. 1972. Continental Aspects of Shrub Distribution, Utilization and Potentials: Africa - The Mediterranean Region. USDA, For. Serv. Gen. Tech. Rep. INT-1, p. 26-36.

McKell, Cyrus M., and Anthony A. Adegbola. 1966. Need for a Range Management Approach for Nigerian Grasslands. J. Range Mgt. 19(6):330-33.

Mahmoud, Moustafa Imam. 1978. Potentialities for Improving Range Management in Mediterranean Coastal Desert of Egypt. Proc. Internat. Rangeland Cong. 1:45-47.

Maloiy, Geoffrey M.O., and Harold F. Heady. 1965. Grazing Conditions in Kenya Masailand. J. Range Mgt. 18(5):269-72.

Marshall, B., D.T. Torell, and R.M. Bredon. 1967. Comparison of Tropical Forages of Known Composition With Samples of These Forages Collected by Esophageal Fistulated Animals. J. Range Mgt. 20(5):310-13.

Mentis, Michael T. 1978. Economically Optimal Species--Mixes and Stocking Rates for Ungulates in South Africa. Proc. Internat. Rangeland Cong. 1:146-49.

Meredith, D. (Ed.). 1955. The Grasses and Pastures of South Africa. Central News Agency, Johannesburg, S. Africa. 771 p.

Moore, Duane G., and Edward J. Britten. 1964. A Comparison of Rhizobium Strains for Effective Nodulation in Kenya Clover, Trifolium semipilosum. J. Range Mgt. 17(6):335-37.

Naveh, Z. 1965. The Importance of the Integrated Sociological and Ecological Approach to the Development of Semi-arid Grasslands in East Africa. Proc. Internat. Grassland Cong. 9:1563-65.

Ngethe, John C. 1976. Preference and Daily Intake of Five East African Grasses by Zebras. J. Range Mgt. 29(6):510.

Ngethe, John C., and Thadis W. Box. 1976. Botanical Composition of Eland and Goat Diets on an Acacia-grassland Community in Kenya. J. Range Mgt. 29(4):290-93.

Obeid, Mubarak M. 1978. The Impact of Human Activities and Lande Use Practices on Grazing Lands in the Sudan. Proc. Internat. Rangeland Cong. 1: 48-51.

O'Rourke, James T. 1978. Grazing Rate and System Trial over Five Years in a Medium-height Grassland of Northern Tanzania. Proc. Internat. Rangeland Cong. 1:563-66.

Oyenuga, V.A., and F.O. Olubajo. 1966. Productivity and Nutritive Value of Tropical Pastures at Ibadan. Proc. Internat. Grassland Cong. 10:962-69.

Pratchett, David. 1978. Effects of Bush Clearing on Grasslands in Botswana. Proc. Internat. Rangeland Cong. 1:667-70.

Pratchett, David, and Bernard Schirvel. 1978. The Testing of Grazing Systems on Semiarid Rangeland in Botswana. Proc. Internat. Rangeland Cong. 1:567-68.

Pratt, D.J., and M.D. Gwynne (Eds.). 1977. Rangeland Management and Ecology in East Africa. Hodder and Stoughton, Ltd., London. 310 p.

Rains, A. Blair. 1970. The Evaluation of African Rangeland Using Aerial Photographs. Proc. Internat. Grassland Cong. 11:99-101.

______. 1978. Milk at the Expense of Meat: The Dilemma of the African Pastoralist. Proc. Internat. Rangeland Cong. 1:123-26.

Robertson, J.H., Gene F. Payne, and C.V. Jensen. 1971. Range Education in East Africa. J. Range Mgt. 24(3):171-74.

Rodel, M.G.W. 1970. Herbage Yields of Five Grasses and Their Ability to Withstand Intensive Grazing. Proc. Internat. Grassland Cong. 11:618-21.

Salih, Mohamed S. 1978. Economics and Problems of Rangeland Productivity in Sudan. Proc. Internat. Rangeland Cong. 1:134-36.

Savory, Allan. 1978. A Holistic Approach to Range Management Using Short Duration Grazing. Proc. Internat. Rangeland Cong. 1:555-57.

Scott, J.D. 1952. The Management of Range Pastures (Veld) in Africa. Proc. Internat. Grassland Cong. 6:477-83.

Shawesh, Othman M., and Herbert G. Fisser. 1978. Ecology of Rangeland Communities in Northwestern Libya, Africa. Proc. Internat. Rangeland Cong. 1:190-92.

Skovlin, Jon M. 1971. Ranching in East Africa: A Case Study. J. Range Mgt. 24(4):263-70

Skovlin, Jon M., and D. Leroy Williamson. 1978. Bush Control and Associated Tsetse Fly Problems of Rangeland Development on the Coastal Plain of East Africa. Proc. Internat. Rangeland Cong. 1:581-83.

Stubbendieck, J. 1978. Constraints to Improvement of Rangeland and Livestock in the Central Plains and Central Plateau of Morocco. Proc. Internat. Rangeland Cong. 1:140-42.

Tainton, Neil M. 1978. Fire in the Management of Humid Grasslands in South Africa. Proc. Internat. Rangeland Cong. 1:684-86.

Talbot, L[ee].M., and L.W. Swift. 1965. Production of Wildlife in Support of Human Populations in Africa. Proc. Internat. Grassland Cong. 9:1355-59.

Talbot, Lee M., and Martha H. Talbot. 1963. The Wildebeest in Western Masailand, East Africa. Wildl. Monogr. 12. 88 p.

Theriez, M., and M. Skouri. 1970. Performance of Sheep in Central Tunisia and Relationship With Diet. Internat. Grassland Cong. 11:767-70.

Theron, E.P., and A.D. Venter. 1978. Methods and Techniques for the Replacement of the Native Grassland in South Africa by Low Cost Techniques. Proc. Internat. Rangeland Cong. 1:620-22.

Trollope, Winston S.W. 1978. Fire--A Rangeland Tool in Southern Africa. Proc. Internat. Rangeland Cong. 1:245-47.

Valenza, J. 1970. Survey of Different Types of Natural Pasture Land in the Senegal Republic. Proc. Internat. Grassland Cong. 11:78-82.

Van Niekerk, J.P., F.V. Bester, and H.P. Lombard. 1978. Control of Bush Encroachment by Aerial Herbicide Spraying. Proc. Internat. Rangeland Cong. 1:659-63.

Van Rensburg, H.J. 1952. Encroachment and Control of Shrubs in Africa in Relation to Grassland Development. Proc. Internat. Grassland Cong. 6:585-91.

______. 1960. Ecological Aspects of the Major Grassland Types in Tanganyika. Proc. Internat. Grassland Cong. 8:367-70.

Van Voorthuizen, E.G. 1970. A Grazing Potential in the Tanga Region of Tanzania. J. Range Mgt. 23(5):325-30.

______. 1971. Cattle Dips Are Used as a Tool for Range Management in Masailand, Tanzania. J. Range Mgt. 24(4):314-15.

______. 1978. Global Desertification and Range Management: An Appraisal. J. Range Mgt. 31(5):378-80.

Wagner, Frederic H. 1978. Tunisian PreSaharan Project: I. Ecological Impact of Cultural Change. Proc. Internat. Rangeland Cong. 1:67-70.

Walker, B. 1970. Utilization of the Natural Pastures on the Hardpan Soils at Ukiriguru, Western Tanzania. Proc. Internat. Grassland Cong. 11:82-85.

Europe

Biswell, Harold H. 1964. Range Management in the General Economy of Greece. J. Range Mgt. 17(6):299-304.

Buzi, Vincent P. Carocci. 1952. Fertirrigation Practices in Italy. J. Range Mgt. 5(4):212-14.

Caputa, J. 1966. Forage Production in Relation to Altitude. Proc. Internat. Grassland Cong. 10:846-51.

Fenley, John M. 1950. Pollarding - Age-old Practice Permits Grazing in Pays Basque Forests. J. Range Mgt. 3(4):316-18.

Fryer, J.D., and S.A. Evans (Eds.). 1968. (5th Ed.). Weed Control Handbook. Volume II. Recommendations. Blackwell Sci. Pub., Oxford, Eng. 325 p.

______. 1970. (5th Ed.). Weed Control Handbook. Volume I. Principles. Blackwell Sci. Pub., Oxford, Eng. 494 p.

Grennan, E.J., and M.A. O'Toole. 1966. Pasture Establishment and Maintenance on Blanket-peat Soil. Proc. Internat. Grassland Cong. 10:842-46.

Gudmundsson, Olafur, Andrew Arnalds, Bjorn Sigurbjornsson, et al. 1978. Experiments on Utilization and Conservation of Grasslands in Iceland. Proc. Internat. Rangeland Cong. 1:576-78.

Hughes, Roy. 1970. Factors Involved in Animal Production from Temperate Pastures. Proc. Internat. Grassland Cong. 11:A31-A38.

Klemme, Marvin. 1955. Water Development as a Prelude to Range Management in Greece. J. Range Mgt. 8(6):271-73.

Liiv, J.G. 1966. Changes in Swards of Natural Grasslands Under the Influence of Fertilization and Utilization. Proc. Internat. Grassland Cong. 10:839-42.

Long, Gilbert A., Michel Etienne, Paule S. Poissonet, and Michel M. Thiault. 1978. Inventory and Evaluation of Range Resources in "Maquis" and "Garrigues" (French Mediterranean Area): Productivity Levels. Proc. Internat. Rangeland Cong. 1:505-9.

Margaropoulos, Panos. 1952. Mountain Range Management and Improvement in Greece. J. Range Mgt. 5(4):200-6.

Papanastasis, Vasilios P. 1978. Potential of Certain Range Species for Improvement of Burned Brushlands in Greece. Proc. Internat. Rangeland Cong. 1:715-17.

Phillips, John, and Robert Moss. 1977. Effects of Subsoil Draining on Heather Moors in Scotland. J. Range Mgt. 30(1):27-29.

Rabotnov, T.A. 1966. An Attempt to Determine the Effect of Different Legumes on Yields and Botanical Composition of Swards in Natural Meadows. Internat. Grassland Cong. 10:835-38.

Scotter, George W. 1965. Reindeer Ranching in Fennoscandia. J. Range Mgt. 18(6):301-5.

Spedding, C.R.W., and E.C. Diekmahns (Eds.). 1972. Grasses and Legumes in British Agriculture. Commonwealth Agricultural Bureau, Slough, Eng. 527 p.

Steen, Eliel. 1966. Investigations Into Reindeer Grazing in North Scandinavia. Proc. Internat. Grassland Cong. 10:998-1003.

Thorsteninsson, I., G. Olafsson, and G.M. VanDyne. 1971. Range Resources of Iceland. J. Range Mgt. 24(2):86-93.

White, W.T. 1950. Pastures in the Italian Highlands. J. Range Mgt. 3(1): 22-28.

Zabello, D.A. 1960. The Influence of Grazing on the Productivity and Change of Botanical Composition of Various Grass Mixtures. Proc. Internat. Grassland Cong. 8:363-65.

Zurn, F. 1966. Methods for Improving the Yields of Meadows in Southern Germany and Austria. Proc. Internat. Grassland Cong. 10:831-34.

Southwestern Asia

Al-Ani, Tariq A., M.M. Al-Mufti, N.A. Ouda, R.N. Kaul, and D.C.P. Thalen. 1974. A Reconnaissance Survey of Range Cover Types in the Western and Southern Deserts of Iraq. Proc. 12th Intern. Grassland Cong., Vol. 1, pt. 2, p. 583-95.

Alinoglue, Nazmi, and Nazim Durlu. 1970. Subterranean Vetch Seed Enhances Persistence Under Grazing and Severe Climates. J. Range Mgt. 23(1): 61-63.

Bhattacharya, A.N., and M. Harb. 1973. Sheep Production on Natural Pasture by Roaming Bedouins in Lebanon. J. Range Mgt. 26(4):266-69.

Boyko, H. 1952. On Regeneration Problems of Destroyed Pasture Areas in Arid Regions. Proc. Internat. Grassland Cong. 6:632-38.

Bryan, Hugh M., and H. Wayne Springfield. 1955. Range Management in Iraq - Findings, Plan and Accomplishment. J. Range Mgt. 8(6):249-56.

Cornelius, Donald R. 1962. Grazing Problems in Turkey. J. Range Mgt. 15(5):257-59.

Draz, Omar. 1978. Revival of the Hema System of Range Reserves as a Basis for Syrian Range Development Program. Proc. Internat. Rangeland Cong. 1: 100-3.

Ellern, S[igmund].J. 1972. Rooting Cuttings of Saltbush (*Atriplex halimus* L.) J. Range Mgt. 25(2):154-55.

Ellern, Sigmund J., and Naphtali H. Tadmor. 1966. Germination of Range Plant Seeds at Fixed Temperatures. J. Range Mgt. 19(6):341-45.

Ellern, Sigmund J., Yochai B. Samish, and David Lachover. 1974. Salt and Oxalic Acid Content of Leaves of the Saltbush *Atriplex halimus* in the Northern Negev. J. Range Mgt. 27(4):267-71.

Eyal, Ezra, Roger W. Benjamin, and Naphtali H. Tadmor. 1975. Sheep Production on Seeded Legumes, Planted Shrubs, and Dryland Grain in a Semiarid Region of Israel. J. Range Mgt. 28(2):100-7.

Gutman, Mario. 1978. Primary Production of Transitional Mediterranean Steppe. Proc. Internat. Rangeland Cong. 1:225-28.

Harlan, Jack R. 1950. Collecting Forage Plants in Turkey. J. Range Mgt. 3(3):213-19.

_____. 1954. Range Management in Turkey. J. Range Mgt. 7(5):220-22.

Heady, Harold F. 1963. Comments on Range Management Technical Assistance in the Middle East With Special Reference to Saudi Arabia. J. Range Mgt. 16(6):317-21.

Jones, D. Kenneth. 1955. Combining Pasture Improvement and Carob Production in Cyprus. J. Range Mgt. 8(4):151-54.

Kernick, M.D. 1970. Ecological Aspects of Range Study and Improvement in the Near East and North Africa. Proc. Internat. Grassland Cong. 11:13-16.

Khan, Ch. M. Anwar. 1968. Sand Dune Rehabilitation in Thal, Pakistan. J. Range Mgt. 21(5):316-21.

_____. 1971. Rainfall Pattern and Monthly Forage Yields in Thal Ranges of Pakistan. J. Range Mgt. 24(1):66-70.

Khan, Muhammad Ihsan-Ur-Rehman. 1952. Improvement and Management of Rangelands in West Pakistan. Proc. Internat. Grassland Cong. 6:491-98.

McArthur, Ian D., and Graham N. Harrington. 1978. A Grazing Ecosystem in Western Afghanistan. Proc. Internat. Rangeland Cong. 1:596-99.

Mariam, Emmanuel K., and James G. Ross. 1972. Intermediate and Pubescent Wheatgrass Complex in Native Collections from Eastern Turkey. Crop Sci. 12(4):472-74.

Miles, Wayne H. 1952. Range Management in Israel, Yesterday, Today, and Tomorrow. J. Range Mgt. 5(4):207-11.

Moghaddam, Mohammad Reza. 1976. Late Fall Vs. Spring Seeding in the Establishment of Crested Wheatgrass in the Zarand Saveh Region of Iran. J. Range Mgt. 29(1):57-59.

_____. 1977. Reseda lutea: A Multipurpose Plant for Arid and Semiarid Lands. J. Range Mgt. 30(1):71-72.

Morag, M. 1972. A Model for the Stratification of Dairy and Mutton Sheep Breeds in Middle Eastern Deserts. J. Range Mgt. 25(4):296-300.

Naveh, A. 1972. The Role of Shrubs and Shrub Ecosystems in Present and Future Mediterranean Land Use. USDA, For. Serv. Gen. Tech. Rep. INT-1, p. 414-27.

Naveh, Z. 1955. Some Aspects of Range Improvement in a Mediterranean Environment. J. Range Mgt. 8(6):265-70.

_____. 1960. Mediterranean Grasslands in California and Israel. J. Range Mgt. 13(6):302-6.

_____. 1970. Effect of Integrated Ecosystem Management on Productivity of a Degraded Mediterranean Hill Pasture in Israel. Proc. Internat. Grassland Cong. 11:59-63.

Naveh, Z., and B. Ron. 1966. Agro-ecological Management of Mediterranean Ecosystems - The Basis for Intensive Pastoral Hill-land Use in Israel. Proc. Internat. Grassland Cong. 10:871-74.

Nemati, Nasser. 1977. Comparative Palatability of Atriplex canescens. J. Range Mgt. 30(5):368-69.

_____. 1977. Range Rehabilitation Problems of the Steppic Zone of Iran. J. Range Mgt. 30(5):339-42.

_____. 1977. Shrub Transplanting for Range Improvement in Iran. J. Range Mgt. 30(2):148-50.

_____. 1978. Range Improvement Practices in Iran. Proc. Internat. Rangeland Cong. 1:631-32.

Ofer, Yitzchak, and No'am G. Seligman. 1969. Fertilization of Annual Range in Northern Israel. J. Range Mgt. 22(5):337-41.

Orev, Yaaqov. 1956. Brush Invasion - 1500 B.C. and 1950 A.D. J. Range Mgt. 9(1):6-7.

Park, Barry C. 1955. Use of Photo Mosaics as a Base for Range Resource Inventory in the Hashemite Kingdom of the Jordan. J. Range Mgt. 8(6):257-60.

Pearse, C. Kenneth. 1970. Range Deterioration in the Middle East. Proc. Internat. Grassland Cong. 11:26-30.

_____. 1971. Grazing in the Middle East: Past, Present and Future. J. Range Mgt. 24(1):13-16.

_____. 1973. Qanats in the Old World: Horizontal Wells in the New. J. Range Mgt. 26(5):320-21.

Peymani, B., and A. Tarifi. 1974. Rehabilitation of Denuded Rangelands Through the Research on Season, Method, Depth, and Planting Rate of Dryland Forage Species. Proc. Intern. Grassland Cong., Vol. 1, Pt. 2, p. 805-12.

Pringle, William L., and Donald R. Cornelius. 1968. Grazing Potential in Aegean Turkey. J. Range Mgt. 21(3):151-54.

Said, Mohammad. 1960. Development of Range Lands in Quetta-Kalat Region, West Pakistan. Proc. Internat. Grassland Cong. 8:220-23.

Seligman, No'am G., and Mario Gutman. 1968. Cattle and Vegetation Responses to Management of Mediterranean Rangeland in Israel. Proc. Internat. Rangeland Cong. 1:616-18.

Tadmor, N[aphtali].H. 1965. Range Development in the Central Negev of Israel. Proc. Internat. Grassland Cong. 9:1443-49.

Tadmor, N[aphtali].H., M. Evenari, and J. Katznelson. 1968. Seeding Annuals and Perennials in Natural Desert Range. J. Range Mgt. 21(5):330-31.

Tadmor, Naphtali H., Ezra Eyal, and Roger W. Benjamin. 1974. Plant and Sheep Production in Semiarid Annual Grassland in Israel. J. Range Mgt. 27(6): 427-33.

Tarman, Omer. 1952. Forage Resources and Ecology of the High-steppes Area of Turkey. Proc. Internat. Grassland Cong. 6:646-49.

Tosun, F., I. Manga, M. Altin, and Y. Serin. 1977. A Study of the Improvement of Dry-land Ranges Developed Under the Ecological Conditions of Erzurum (Eastern Anatolia). Proc. 13th Intern. Grassland Cong., Sec. 3-5, p. 249-55.

USSR and East Asia

Agababyan, Sh. 1966. Alpine Grasslands of the Armenian SSR and Their Utilization and Improvement. Proc. Internat. Grassland Cong. 10:860-63.

Andreyev, N. 1974. Potentialities of Native Haylands and Pastures in the Soviet Union. Proc. 12th Intern. Grassland Cong., Vol. I, Pt. I, p. 165-75.

Banerji, J. 1952. Modern Trends in Ecological Management of Grasslands in the World Background With Special Reference to the Tropics. Proc. Internat. Grassland Cong. 6:513-20.

Bharucha, F.R., and R.O. Whyte. 1952. The Grazing and Fodder Resources of India. Proc. Internat. Grassland Cong. 6:1446-51.

Dabadghao, P.M. 1960. Types of Grass Covers of India and Their Management. Proc. Internat. Grassland Cong. 8:226-30.

Dabadghaeo, P.M., and K.A. Shankarnarayan. 1970. Studies of *Iseilema*, *Sehima*, and *Heteropogon* Communities of the *Sehima-Dichanthium* Zone. Proc. Internat. Grassland Cong. 11:36-38.

Dakshini, K.M.M. 1972. Continental Aspects of Shrub Distribution, Utilization, and Potentials: Indian Subcontinent. USDA, For. Serv. Gen. Tech. Rep. INT-1, p. 3-15.

Dutton, W.L. 1952. Management and Administration of Range Lands in Japan. J. Range Mgt. 5(4):221-29.

Gorrie, R. Maclagan. 1952. Land Use, Soil Erosion, and Livestock Problems in Ceylon. J. Range Mgt. 5(4):215-20.

Gupta, R.K., and P.C. Nanda. 1970. Grassland Types and Their Ecological Succession in the Western Himalayas. Proc. Internat. Grassland Cong. 11:10-13.

Han, I.K., D.A. Kim, and S.H. Park. 1970. Seasonal Changes in Chemical Composition and of Dry Matter Digestibility of Korean Native Herbage Plants. Proc. Internat. Grassland Cong. 11:92-95.

Ibragimov, I. 1974. Establishment and Utilization of Cultivated Pastures in Desert Zone. Proc. 12th Intern. Grassland Cong. Vol. 1, Pt. 2, p. 734-38.

Krishnamurthy, L. 1978. Herbage Biomass Changes on Some Indian Rangelands. Proc. Internat. Rangeland Cong. 1:221-24.

Larin, I.V. 1960. Grassland Management in the U.S.S.R. Proc. Internat. Grassland Cong. 8:356-60.

_____. 1966. The Peculiarities of Pasture Utilization in the Different Climatic Zones of the USSR. Proc. Internat. Grassland Cong. 10:823-28.

Lekborashvili, G. 1974. Effectiveness of Different Methods of Improving High Mountain Pastures. Proc. 12th Intern. Grassland Cong., Vol. 1, Pt. 2, p. 749-55.

Nechaeva, Nina T. 1960. Rotation of the Pastures in the Desert of Turkmen S.S.R. Proc. Internat. Grassland Cong. 8:360-62.

_____. 1965. Sown Winter Ranges in the Middle Asia Foothill Deserts. Proc. Internat. Grassland Cong. 9:1365-67.

_____. 1974. Phytomass Structure and Productivity of Central Asian Desert Pastures in Relation to Vegetation Types. Proc. 12th Intern. Grassland Cong., Vol. 1, Pt. 1, p. 122-32.

Oohara, H., N. Yoshida, K. Fukunaga, et al. 1974. Productivity of Public

Grassland Established at High Elevation in Hokkaido, Northern Japan. Proc. 12th Intern. Grassland Cong., Vol. 1, Pt. 2, p. 790-804.

Pandeya, S.C., N.R. Mankad, and H.K. Jain. 1974. Potentialities of Net Primary Production of Arid and Semiarid Grazinglands of India. Proc. 12th Intern. Grassland Cong., Vol. 1, Pt. 1, p. 133-64.

Pearse, C. Kenneth. 1965. Range Study Tour of the Soviet Union. J. Range Mgt. 18(6):305-10.

Pearson, Henry A. 1975. An Agricultural Tour of the Soviet Union. Rangeman's J. 2(6):181-83.

Petrov, M.P. 1972. Continental Aspects of Shrub Distribution, Utilization and Potentials: Asia. USDA, For. Serv. Gen. Tech. Rep. INT-1, p. 37-50.

Rusin, V. 1974. Some Aspects of Vertical Zonality of Mountain Vegetation in Central Caucasus. Proc. 12th Intern. Grassland Cong., Vol. 3, Pt. 1, p. 455-59.

Sajor, Valentin, and Teofilo Santos. 1952. Grasses in the Philippines. Proc. Internat. Grassland Cong. 6:1452-58.

Sant, Harshwardhan R. 1964. Grazing Susceptibility Numbers of Grasses and Forbs from the Grazing Grounds of Varansi, India. J. Range Mgt. 17(6):337-39.

_____. 1964. Seasonal Variation in Coverage of Selected Grasses and Forbs in Relation to Grazing Intensities in India. J. Range Mgt. 17(2):74-76.

_____. 1966. Grazing Effects on Grassland Soils of Varanasi, India. J. Range Mgt. 19(6):362-67.

Shamsoutdinov, I. Sh. 1966. The Improvement of Desert Ranges in Uzbekistan. Proc. Internat. Grassland Cong. 10:960-62.

Shamsutdinov, Z. 1970. Creation of Permanent Pastures in the Uzbekistan Desert Zone. Proc. Internat. Grassland Cong. 11:69-71.

Sinkovsky, L. 1974. Increasing Forage Production in Semisavannah Ranges in Central Asia's South. Proc. 12th Intern. Grassland Cong., Vol. 1, Pt. 2, p. 841-44.

Smurygin, M. 1974. Basic Trends of Grassland Research in the USSR. Proc. 12th Intern. Grassland Cong., Vol. 1, Pt. 1, p. 76-88.

Tamesis, Florencio, and Valentin Sajor. 1952. Forest Grazing in the Philippines. Internat. Grassland Cong. 6:1556-60.

Walandow, P.H., and S. Bone. 1952. Grassland in Indonesia. Proc. Internat. Grassland Cong. 6:1498-1503.

Zhambakin, Zh., S. Pryanishnikov, K. Baitkanov, and P. Salukov. 1974. Improvement and Rational Utilization of Native Grasslands in Kazakhstan. Proc. 12th Intern. Grassland Cong., Vol. 1, Pt. 2, p. 884-89.

Appendix

SUGGESTED OUTLINE FOR A RANGE SCIENCE LITERATURE SEARCH SEMINAR

Most freshmen and sophomores receive a general introduction to the university library, but formalized training in literature searching in their selected professional disciplines is seldom provided. Undergraduate range science students need such training as soon as they have designated their academic major and begin taking courses in their major field, preferably as juniors or even second-semester sophomores. Even more intensive literature search training is needed by graduate students.

A literature search seminar is one way of providing such training at the sophomore-junior or graduate student level. This guide has been prepared to serve as a syllabus to a literature search seminar in range science. The primary objectives of the seminar would be to survey the range science discipline, relate literature searching to the discipline, and project professional improvement needs. A one semester hour seminar for an undergraduate seminar or a one or two semester hour seminar for graduate students is suggested. Such seminars will generally be in addition to the typical senior range science seminar in which a review of selected current literature, recent developments in the field of range science, or professional opportunities or a combination of these are emphasized.

Through the use of this manual, the range science instructor will be able to teach a literature search seminar. However, assistance should be sought when needed from the agricultural or science reference librarian at the university. Part or all of the sessions might be held in the library with ready access to appropriate reference sections. It is suggested that each student's participation in the seminar be enhanced by making an exhaustive literature search on a selected topic, this resulting in a well-edited review paper or abstract bibliography.

The following outline, including 16 sessions, is suggested for a one semester hour range science literature search seminar:[51]

1. Introduction to range science; study p. 1-5.
2. Abstracting and extracting procedures; use of CBE Style Manual suggested for making bibliographic citations; also select and refine research topic.
3. Literature searching procedures; study p. xi, 7-11.
4. Library classification systems; study p. 11-17.

5. Indexing and abstracting literature; study p. 17-23.
6. Indexing and abstracting literature (continuation)
7. Computerized literature retrieval; study p. 24-32.
8. Current awareness and professional development; study p. 32-34.
9. Literature used by range scientists; study p. 35-36.
10. Major periodicals and serials; study p. 36-40.
11. Related periodicals and serials; study p. 40-48.
12. Agencies and organizations as information sources; study p. 61-86.
13. Developing a personal library.
14. Classroom presentation of student topics. (Note: presentation of student topics could be extended to sessions 13 and 16 if needed.)
15. Classroom presentation of student topics.
16. Final summary and critique.

NOTES

[1]Society for Range Management. 1972. Benchmarks: A Statement of Concepts and Positions. SRM, Denver, Colo. 21 p.

[2]Vallentine, John F. 1978. More Pasture or Just Range for Rangemen? Rangeman's J. 5(2):37-38.

[3]USDA, Forest-Range Task Force. 1972. The Nation's Range Resources--A Forest-Range Environmental Study. USDA, For. Serv. For. Resource Rep. 19. 147 p.

[4]Ibid, Society for Range Management, 1972.

[5]Stoddart, Laurence A., Arthur D. Smith, and Thadis W. Box. 1975 (3rd Ed.). RANGE MANAGEMENT. McGraw-Hill Book Co., New York, p. 3.

[6]A four-level subject classification system for range science can be found in the following: Vallentine, John F. 1978. U.S.-CANADIAN RANGE MANAGEMENT, 1935-1977: A SELECTED BIBLIOGRAPHY ON RANGES, PASTURES, WILDLIFE, LIVESTOCK, AND RANCHING. Oryx Press, Phoenix, Ariz. 337 + 17 p. (8 1/2 x 11 in.).

[7]This section was in part abbreviated from the following: (1) Campbell, Robert S. 1948. Milestones in Range Management. J. Range Mgt. 1(1):4-8; (2) Chapline, William R., Robert S. Campbell, Raymond Price, and George Stewart. 1944. The History of Western Range Research. Agric. Hist. 18(3): 127-43; (3) Pechanec, Joseph F. 1957. The History and Accomplishments of Our Range Society. J. Range Mgt. 10(4):189-93; (4) Sampson, Arthur W. 1954. The Education of Range Managers. J. Range Mgt. 7(5):207-12; (5) Stoddart, Laurence A. 1950. Range Management. Pages 113-35 in Robert K. Winters (Ed.). FIFTY YEARS OF FORESTRY IN THE U.S.A. Society of American Foresters, Washington, D.C.; and (6) Stoddart, Laurence A., Arthur D. Smith, and Thadis W. Box. 1975 (3rd Ed.). RANGE MANAGEMENT. McGraw-Hill Book Co., New York, p. 76-77.

[8]Skovlin, Jon M. 1976. References to "Range." Rangeman's News 3(2):5.

[9]Jardine, James T., and Mark Anderson. 1919. Range Management on the National Forests. USDA Bul. 790. 98 p.

[10]Sampson, Arthur W. 1923. RANGE AND PASTURE MANAGEMENT. John Wiley & Sons, New York. 421 p.; Sampson, Arthur W. 1928. LIVESTOCK HUSBANDRY ON RANGE AND PASTURE. John Wiley & Sons, New York. 411 p.

[11]Stoddart, Laurence A., and Arthur D. Smith. 1943. (1st Ed.). RANGE MANAGEMENT. McGraw-Hill Book Co., New York. 547 p.

[12]Further assistance in terminology clarification and equivalence can be found in the following: (1) USDA, Natl. Agric. Lib. 1963. SUBJECT HEADING LIST, 4 vols. USDA, Natl. Agric. Lib., Washington, D.C. 2,434 p.; (2) U.S. Lib. of Cong., Subject Cataloging Div. 1975 (8th Ed.). LIBRARY OF CONGRESS HEADINGS, 2 vols. Library of Cong., Washington, D.C. 2,026 p.; and (3) Oryx Press. 1978 (2nd Ed.). AGRICULTURAL TERMS. Oryx Press, Phoenix, Ariz. 122 p. (Its vocabulary is the source for the headings appearing in the subject index of the Bibliography of Agriculture.)

[13]Dewey, Melvil (Devised By). 1971 (18th Ed.). DEWEY DECIMAL CLASSIFICATION AND RELATIVE INDEX, 3 vols. Lake Placid Club, New York City. 2,692 p. (Volume 1, introduction and standard subdivision tables; Volume 2, schedules of the Dewey Decimal Systems [details of the classification system]; and Volume 3, the relative index).

[14]Additional information about the Library of Congress Classification System can be found in the following: U.S. Library of Congress, Subject Cataloging Division. 1975 (3rd Ed.). LC CLASSIFICATION OUTLINE. U.S. Govt. Print. Office, Washington, D.C.; 26 p.; and Immroth, John Phillip. 1971 (2nd Ed.). A GUIDE TO THE LIBRARY OF CONGRESS CLASSIFICATION SYSTEM. Libraries Unlimited, Littleton, Colo. 335 p.

[15]See also: Morehead, Joe. c1975. INTRODUCTION TO UNITED STATES PUBLIC DOCUMENTS. Libraries Unlimited, Littleton, Colo.; 289 p.

[16]MONTHLY CATALOG OF U.S. GOVERNMENT PUBLICATIONS. U.S. Supt. of Doc., Washington, D.C. (A monthly publication beginning in 1941).

[17]Bourne, Charles P. 1969. Overlapping Coverage of Bibliography of Agriculture by 15 other Secondary Services. Information General Corp., Palo Alto, Calif. (Prepared for the National Agricultural Library; reproduced by Clearinghouse, U.S. Dept. of Commerce.) Based on 5000 citations sampled uniformly from the 1967 issues of Bibliog. of Agric., only 46 percent of its

citations were covered by the other 15 abstracting and indexing publications. None of the 15 services overlapped more than 20 percent of the Bibliog. of Agric. data base, but some of the Bibliog. of Agric. citations were covered by as many as six of the 15 other secondary services.

[18]USDA. 1977. USDA Data Base Directory. Automated Data Systems. USDA, Washington, D.C.

[19]Caponio, Joseph F., and Marilyn C. Broken. 1973. Selected Food and Agricultural Data Bases in the U.S.A. National Agricultural Library, USDA, Washington, D.C.

[20]USDA Brochure, AGRICOLA. Reference Division, National Agricultural Library, Beltsville, Md. 20705.

[21]Ibid, Caponio and Broken, 1973.

[22]Library of Congress. No date. SCORPIO, Information Systems Office, Administrative Department, Library of Congress, Washington, D.C.

[23]Lockheed Information Systems. 1977. The Dialog Service. Lockheed Missles and Space Company, Inc., Palo Alto, Calif. 94304.

[24]System Development Corporation. 1975. SDS Search Service, We Have What You're Looking For. System Development Corporation, 2500 Colorado Avenue, Santa Monica, Calif. 90406.

[25]Vallentine, John F. 1978. U.S.-CANADIAN RANGE MANAGEMENT, 1935-1977: A SELECTED BIBLIOGRAPHY ON RANGES, PASTURES, WILDLIFE, LIVESTOCK, AND RANCHING. Oryx Press, Phoenix, Ariz. 337 + 17 p. (8 1/2 x 11 in.).

[26]Renner, F.G., Edward C. Crafts, Theo. C. Hartman, and Lincoln Ellison. 1938. A SELECTED BIBLIOGRAPHY ON MANAGEMENT OF WESTERN RANGES, LIVESTOCK, AND WILDLIFE. USDA Misc. Pub. 281. 468 p.

[27]Vallentine, John F. 1979. The Literature of Range Science Based on Citations in the Journal of Range Management. Journal of Range Management 32(3):241-43.

[28]These glossaries are recommended for clarifying, understanding, and properly defining terminology (1) unique to range science and (2) commonly used both in range science and related fields; their use is also suggested in developing lists of synonyms and related terms for literature searching.

[29]An annotated bibliography of 58 selected range science bibliographies previously published in monographic or journal article form; these bibliographies vary from narrow topics to broad aspects of range management.

[30]General Services Administration. 1976. United States Government Organization Manual. Office of the Federal Register, National Archives and Records Service, Washington, D.C. 20408.

[31]USDA. 1978. Agriculture USA. Office of Governmental and Public Affairs, USDA, Washington, D.C. 20250.

[32]USDA. 1977. Your United States Department of Agriculture, How It Serves People On the Farms and in the Community, Nation, and World. USDA PA-824 (Office of Communications, USDA, Washington, D.C. 20250).

[33]USDA. 1976. Fact Book of U.S. Agriculture. USDA Misc. Pub. 1063 (Office of Communication, USDA, Washington, D.C. 20250).

[34]USDA. 1977. Popular Publications for the Farmer, Surburbanite, Homemaker, Consumer. List No. 5, Publications Division, Office of Communication, USDA, Washington, D.C. 20250.

[35]USDA. 1976. List of Available Publications of the United States Department of Agriculture. List No. 11, Publications Division, Office of Communication, USDA, Washington, D.C. 20250.

[36]Cooperative State Research Service. 1976. The States Role in the Cooperative State-Federal Research System for Agriculture and Forestry. USDA Agric. Infor. Bul. 288.

[37]SCS. 1976. National Range Handbook: Rangeland, Grazable Woodland, and Native Pasture. USDA, SCS, Washington, D.C. 20250.

[38]SCS. 1962. Classifying Rangeland for Conservation Planning. USDA Agric. Handbook 235 (USDA, SCS, Washington, D.C. 20250).

[39]SCS. 1978. Range Conservationist. USDA, SCS, Washington, D.C. 20250.

[40]SCS. 1976. Assistance Available From the Soil Conservation Service. USDA Agric. Info. Bul. 345 (USDA, SCS, Washington, D.C. 20250).

[41]Ibid, footnote no. 30.

[42]Canada Department of Agriculture. 1975. Organization and Activities of the Canada Department of Agriculture. Canada Department of Agriculture Pub. 1123 (Ottawa, Canada K1A 0C7).

[43]Canada Department of Agriculture. 1978. Information Division at Your Service. Information Division. Canada Department of Agriculture, Ottawa, Canada K1A 0C7.

[44]Canada Department of Agriculture. 1974. Guide to Federal Agricultural Services. Information Division, Canada Department of Agriculture Pub. 1262 (Ottawa, Canada K1A 0C7).

[45]Canada Department of Agriculture. 1978. Publications for Farm and Home. Canada Department of Agriculture Pub. 5103. (Ottawa, Canada K1A 0C7).

[46]Canada Department of Agriculture. 1967. Canada Agriculture, the First Hundred Years. Canada Department of Agriculture Historical Series, No. 1 (Ottawa, Canada K1A 0C7).

[47]A highly selected, annotated bibliography including 35 literature items recommended for background reading in range science; these were selected to represent the hard core of range science while providing lead-in to those phases common to related academic areas as well.

[48]Selected annotated bibliography of about 500 key range science literature limited to U.S. and Canadian sources; references have been arranged to represent and emphasize the principle phases or component parts of range science; selections were limited to monographic literature based on its less frequent use and generally broader subject matter coverage per item than journal literature.

[49]This section emphasizes big game and predators; upland game birds, waterfowl, and most non-game wildlife have been excluded.

[50]A selected list of about 400 references providing insight into the application of range science outside the U.S. and Canada; selections are restricted to English-language literature generally available in major U.S. agricultural libraries and include both periodical and monographic literature; planned use is primarily by North Americans desiring an introduction to range management in foreign countries.

[51]For possible library activity assignments to enhance these sessions, the GUIDE TO THE LITERATURE OF THE LIFE SCIENCES (by Roger C. Smith and W. Malcolm Reid; 1972, 8th Ed.; Burgess Pub. Co., Minneapolis, Minn.) or GUIDE TO THE LITERATURE OF THE ZOOLOGICAL SCIENCES (by Roger C. Smith and Reginald H. Painter; 1966, 7th Ed.; Burgess Pub. Co., Minneapolis, Minn.) are suggested.

AUTHOR INDEX

In addition to authors, this index includes editors and compilers. Alphabetization is letter by letter and reference numbers are to page numbers.

A

B

C

D

E

F

G

H

I

J

M

S

T

U

V

INDEX OF ORGANIZATIONS, AGENCIES, AND SERVICES

This index includes all organizations and associations, agencies, services, and data bases. Underlined numbers refer to major entries. Alphabetization is letter by letter and references are to page numbers.

Index of Organizations, Agencies, & Services

I

L

N

O

P

R

S

T

U

W

INDEX OF PERIODICALS, SERIALS, ABSTRACTS, AND BIBLIOGRAPHIES

This index includes titles, periodicals, serials, abstracts, and bibliographies. Underlined numbers refer to main entries. Alphabetization is letter by letter and references are to page numbers.

A

SUBJECT INDEX

This index includes major topics covered in the text. Underlined numbers refer to subjects covered in depth, including chapter and section headings. Literature on foreign countries has been indexed geographically. References are to page numbers and alphabetization is letter by letter.

A

B

H

I

J

K

L

S